ENVIRONMENTAL HAZARDS
Human and Policy Dimensions

ENVIRONMENTAL HAZARDS: Human and Policy Dimensions (EHAZ) is an innovative, interdisciplinary and international research journal addressing the human and policy dimensions of hazards.

The journal addresses the full range of hazardous events from extreme geological, hydrological atmospheric and biological events such as earthquakes, floods, storms and epidemics to technological failures and malfunctions such as industrial explosions, fires and toxic material releases. EHAZ is the source of the new ideas in hazards and risk research.

With a genuinely international perspective, this journal highlights issues of human exposure, vulnerability, awareness, response and risk. The role of hazards in development, and issues of efficiency, social justice and sustainability are also explored in the journal.

Well known conventional hazards receive extensive coverage but submissions about new forms of hazard, emerging risk management institutions and restructuring of ideas about hazards – including their role in human affairs – are particularly welcome.

Reinvigorating the debate about how we define, understand and manage hazards, the journal is interdisciplinary in scope and open to contributions by specialists from a wide range of fields who are interested in the effects of hazards events on people, property and societies.

ENVIRONMENTAL HAZARDS
Human and Policy Dimensions

VOLUME 10 ISSUE 1 2011
SPECIAL ISSUE

Special issue: Reducing disaster risks: Progress and challenges in the Caribbean region

Environmental Hazards 10(1) March 2011.

Published by Earthscan: Dunstan House, 14a St Cross Street, London EC1N 8XA, UK.

ISSN: 1747-7891 (print), 1878-0059 (online)
ISBN: 978-1-84971-357-3

Abstracting services which cover this title include Elsevier Scopus and GeoRef

Printed in the UK by MPG Books Ltd on FSC certified paper.

Environmental Hazards is published quarterly. Periodicals Postage Paid at Rahway, NJ. US agent: Mercury International, 365 Blair Road, Avenel, NJ 07001. POSTMASTER: Address changes to ENVIRONMENTAL HAZARDS, 365 Blair Road, Avenel, NJ 07001.

MIX
Paper from responsible sources
FSC
www.fsc.org
FSC® C018575

earthscan
www.earthscan.co.uk

DEM DEUTSCHEN VOLKE

ENVIRONMENTAL
HAZARDS
Human and Policy Dimensions

Introduction

FRANKLIN J. MCDONALD[1,2,*]

[1]Disaster & Emergency Management Program, School of Administrative Studies, Room 204A, Atkinson Faculty of Liberal and Professional Studies, York University, 4700 Keele Street, Toronto, Ontario, Canada M3J 1P3
[2]Disaster Risk Reduction Centre, University of the West Indies, Mona, Jamaica

In December 2009, the Fourth Annual Caribbean Disaster Management Conference (CDM 4) was held in Montego Bay, Jamaica; it brought together after 25 years a group of persons who had previously participated in the 1984 International Conference on Hazard Mitigation that had taken place in Ocho Rios, Jamaica. Their rich discussions, dialogue and debate on progress made or not made; good, best and poor practice; and lessons learnt, unlearnt and still to be learnt have led to this volume.

A number of participants of that group have now collaborated in providing their thoughts and reflections on developments in the disaster management, risk reduction and hazard studies cluster over the last quarter-century. They have indulged in and shared some crystal-gazing into the challenges still awaiting resolution if human societies are to be safer, settlements more robust and economies more resilient in the coming decades.

Within a month of the December 2009 Conference in Jamaica, the tragedy of the Port au Prince earthquake in Jamaica's close neighbour Haiti began to unfold – illustrating the need for renewed efforts to build on the good practices of the past and to address the complex linkages between risk reduction, governance, environmental degradation, poorly managed urbanization, poverty and capacity building at all levels of government.

The authors' collection of observations, analysis, reflections, conclusions, speculations, opinions and recommendations presents a challenging composite of what has been achieved and what remains to be done!

These reflections seek to identify in an unvarnished fashion the questions that should be answered if progress is to be made in moving societies on to a more sustainable path, achieve resilience, institutionalize risk reduction and improve public safety. There is no pretension here that all questions are identified or answered, but this collection certainly offers a wide and hopefully stimulating range of issues and a framework for grasping the past lessons, tackling the present challenges and building the capacity for future success in disaster risk reduction beyond the Hyogo Framework for Action (HFA) 2015 aspirational date.

Broad consensus emerges in these papers on the wide and diverse range of activities undertaken at the global, regional, national, subnational and local level of governance in many sectors and across disciplines to incorporate risk-related coping and adaptation strategies appropriate to the Caribbean and other regions. Many activities related to disaster risk reduction of course preceded the 1984 event, and in the intervening 25 years particularly under the aegis of the International Decade for Natural Disaster

■ *Email: franklin.jmcd@gmail.com

doi:10.3763/ehaz.2011.0001 ISSN: 1747-7891 (print), 1878-0059 (online) www.earthscan.co.uk/journals/ehaz
earthscan

Reduction and its successor the International Strategy for Disaster Reduction (ISDR). All the Caribbean jurisdictions have supported the HFA targets promoted by the ISDR, although the level of actual implementation of HFA 'platforms' may leave room for improvement.

Of particular relevance in the Caribbean region has been the emergence of the comprehensive disaster management (CDM) framework as the basis for national DRR mainstreaming, as the CDM and other initiatives have been significantly facilitated by the initiation and implementation of several regional cooperative mechanisms including but not confined to those originally created to deal with response, emergency management and preparedness such as Caribbean Disaster Emergency Response Agency (CDERA) (its successor Caribbean Disaster and Emergency Management Agency (CDEMA) now having a clear risk reduction mandate).

The participants in the 1984 Ocho Rios Conference were drawn from many backgrounds and jurisdictions and many have gone on to distinguished careers in risk-related areas of endeavour in the public and private sectors, academia and the professions; they therefore bring rich and diverse experiences, and insights into many facets of a challenging and largely under appreciated area! These reflections by this group of authors thus represent a lengthy span of human experience worth entraining in the current conversation on disaster risk reduction (DRR) and the HFA. These reflections are timely as we tackle afresh the old issues of catastrophic human losses such as those associated with the 2004 Indian Ocean tsunami and the 2010 Haiti earthquake. The death toll in both these events exceeded 200,000 and there is wide consensus that many of these deaths were avoidable and attributable to the lack of application of existing knowledge and available technology, that is tsunami early-warning systems and aseismic construction techniques, respectively.

It is hoped that this volume will facilitate the process of making safer societies in the Caribbean and the wider world, but as the readers will note it does not offer any simple quick or easy solutions to the challenge of building safe, robust and resilient communities, cities or countries. The authors have again aspired – as their earlier work of a quarter-century ago also did – to critically review the state of technical and scientific knowledge and practice. They go beyond the physical dimensions in this volume to focus more attention on governance, political will, capacity and the social dimensions of the risk issues, and they generously share facts, cases, insights and educated opinions. They also go further than the 1984 closing session suggestion by taking note of the role of champions as possible catalysts and facilitators for taking the conversation or debate on risk reduction further afield and to push for and promote implementation and active adaptation in all spheres.

The contributions to this volume build upon the observations of the original Ocho Rios Conference of 1984 and reflect upon developments of the past 25 years. The first three chapters focus on the Caribbean region and aspects of the implementation of CDM in the region. The opening paper was prepared by Jeremy Collymore, Executive Director of the CDEMA, the leading disaster management organization of the region. Mr Collymore initiated his distinguished career in disaster management at the Ocho Rios Conference in 1984. In this paper he recounts the evolution of the regional disaster management approach in the Caribbean over the past 25 years. This development has occurred in parallel with the International Decade for Disaster Reduction and subsequent international initiatives of the ISDR to improve disaster response and to reduce disaster risk. The paper by Keith Ford provides a report on the Caribbean Disaster Mitigation Project of the 1980s that provided a foundation for subsequent regional mitigation efforts. Eleanor Jones provides an in-depth analysis of the progress of disaster management in a single national disaster management organization. Together, these three chapters give a sense of the relationship of disaster management from the project level to the national level and the regional level.

Later topics turn to more general themes. Andrew Maskrey, another graduate of Ocho Rios (1984) who has emerged as a leader in the field, recounts the development of community-based disaster risk management as a global movement. This is a concept that was foreseen in 1984 and has since become a major emphasis for many government and non-government disaster management agencies. Krimgold focuses attention on the critical role of construction quality and regulatory support in reducing vulnerability to natural phenomena. Building and land-use regulation are a fundamental part of governance and public responsibility and a prerequisite for consistent economic and social development.

Richard Olsen and his co-authors address issues of public responsibility more broadly in laying out the steps for creating and maintaining political will for disaster mitigation. Stephen Bender, with the perspective of the last 25 years, assesses the emergence of an independent field of disaster management distinct from development and raises questions. In the final paper Ian Davis provides personal and professional observations from a rich and varied career in disaster risk reduction including the period 1984–2010.

This modest effort at collective stocktaking in the field of disaster risk reduction may be of interest in attempting to measure progress over time. In spite of much hard work and the best of intentions, we have seen vulnerability continue to expand as new hazards emerge. The continued growth of disaster losses calls for rededication and more imaginative and energetic solutions to the problems of living with risk.

ENVIRONMENTAL
HAZARDS
Human and Policy Dimensions

Disaster management in the Caribbean: Perspectives on institutional capacity reform and development

JEREMY COLLYMORE*

Caribbean Disaster and Emergency Management Agency, Building No. 1, Manor Lodge Complex, Lodge Hill, Saint Michael, Barbados, West Indies

This paper is a reflective contribution on the issues and processes that have influenced and informed the development and institutionalization of disaster management practice and policy in the Caribbean over the last two and a half decades. It takes the viewpoint of a 'participant observer' and seeks to not only describe the events and triggers but also to raise some key questions necessary to chart the way forward.

Keywords: Caribbean; disaster management; mitigation; preparedness; recovery; response

1. Introduction

In the last 25 years, in the Caribbean there has emerged an agenda to develop and maintain an institutional framework and capacity for advancing disaster preparedness and to a lesser extent mitigation. The legal status of the national focal points leading these efforts and their resourcing still present many opportunities for enhancement.

The disconnection between our hazard loss experiences and our disaster initiatives suggests that few governments recognize the long-term economic implications of disasters, especially on their debt structure and the role that hazard mitigation can play in easing economic problems (Collymore, 2008; IDB, 2000; Rasmussen, 2004).

In many situations, it is noted that the passing of regulations and development of technology have not impacted loss reduction behaviour and practice. It may be suggested that this in part reflects the need for more attention to be paid to how one can alter the structure of the market as a medium for promoting desired behaviour and practice (Cummins and Mahul, 2010; Collymore, 2008).

Disaster management has become a policy problem of global scope, which requires the examination of ways in which modern developments magnify the vulnerability of human communities to natural and other hazards, and also the opportunities for reducing this (Cummins and Mahul, 2010; UNDP, 2004).

It calls for a conscious policy that recognizes and addresses the link between land-use decisions and settlement policy and the dramatic increase in hazard exposure (UNDP, 1991).

Current and future disaster management agendas must also face the challenges that disaster reduction measures must meet in the face of unmet capital investment needs, demands for more extensive and efficient cost recovery and shifting focus of donar financial support.

Linking disaster management to development and the environment increases the range of stakeholders engaged in the policy discourse and will require some adjustments in the institutional arrangements. Partnering with other programmes in climate change, sea-level rise, poverty reduction and public sector reform can create the synergies necessary to get the holistic

* *Email:* jc@caribsurf.com

doi:10.3763/ehaz.2011.0002 ISSN: 1747-7891 (print), 1878-0059 (online) www.earthscan.co.uk/journals/ehaz

and integrated approach necessary for disaster reduction.

This paper is a reflective contribution on the issues and processes that have influenced and informed the development and institutionalization of disaster management practice and policy in the Caribbean over the last two and a half decades. It takes the viewpoint of a 'participant observer' and seeks to not only describe the events and triggers but also to raise some key questions necessary to chart the way forward.

The reflections are focused primarily on the experiences of 16 English-speaking states of the Caribbean Disaster Emergency Response Agency up to August 2009, now CDEMA (see www.cdema.org).

The ideas I am presenting have been drawn from a number of 'thought pieces' that I have put together over the last 25 years and that are fashioned in a way that provides a general vision, my personal vision, of what is required to make disaster management a centrepiece of the decision-making and policy-forming environment in these small island developing states.

The essence of this reflection is driven by a general overview of where we were, where we need to go and how we can get there. These thoughts are set against the four assertions below:

1. Disaster management in the past was event-driven, that is, activities, programmes and related interventions were largely responsive in nature.
2. Except for recent times, contemporary disaster management has been mono-focused, externally driven and response-oriented.
3. The future of disaster management in our islands will largely be influenced by global economic political consideration in our islands.
4. To move forward towards the comprehensive consolidated framework, we need to make fundamental changes to the institutional arrangements for disaster management at the country level.

2. Overview

A cursory review of disaster management initiatives, programmes, policies and activities within the last 25 years in the region clearly indicates a strong link to the major event impact episodes in the region. I have chosen the last 25 years because the larger part of our generation has a distinct connection and some experience, even though limited, with this period. It also represents a period in which most of the states, with the exception of the Overseas Territories, through independence, became responsible for managing their systems of government and institutions. This brought a cohort of indigenous leaders to the forefront of national decision making and the attendant challenges of making tough choices on development options in the face of limited resources.

2.1. Evidence of exposure and impact

The Caribbean Region, largely comprised of small island developing states (SIDS) and low-lying coastal states (LLCS), is widely acknowledged as being the second most prone to a range of natural hazards, including hurricanes, floods, landslides and occasional volcanic eruptions. The region experiences regular annual losses due to natural hazard events in the order of US$3 billion. More than 68 per cent of these losses are in the social and productive sector. Additionally 60 per cent of the population of the region resides in the coastal zone. Further, 70 per cent of all economic activity takes place within two miles of the coastline (Table 1).

Social and economic losses from disasters in the Caribbean are often severe, including loss of life, loss of homes and livelihoods, environmental damage, damage to critical infrastructure and damage to the economic base of countries. Tourism and agriculture, which form the basis of economic development for most Caribbean states and territories, are typically immediately disrupted when disasters occur.

TABLE 1 Examples of the impact of disaster events on the GDP of Caribbean countries

Time	Place	Disaster	Impact
1988	Jamaica	Hurricane Gilbert	65% GDP
1989	5 countries	Hurricane Hugo	US$412m
1999	Dominica	Hurricane Lenny	53% GDP
2004	Haiti & DR	Flooding	270 deaths
2004	Grenada	Hurricane Ivan	US$895m
2004	Jamaica	Hurricane Ivan	US$592m
2005	Guyana	Flooding	60% GDP

Climate change is likely to affect the Caribbean region's experiences with disasters. This can already be observed in the increasing frequency and severity of extreme weather events experienced. From 1970 to 1979, there were 15 category 3, 4 and 5 hurricanes. From 1980 to 1989, there were 17 category 3, 4 and 5 hurricanes. From 1990 to 1999, there were 25 category 3, 4 and 5 hurricanes experienced in the region. Between 2000 and 2008, there have been more than 29 severe hurricanes in the region. The key issue that is raised by this observation is the need for further enhancement of preparedness and response and mitigation capabilities within the region in light of the anticipated effects of climate change (Table 2).

Existing data also indicate an increase in the number of deaths in the region attributed to earthquakes and tsunamis between 1500 and 1999. Table 1 shows that approximately 16,500 earthquake fatalities occurred since 1600. The most fatal event – a magnitude 8 thrust earthquake – occurred in 1843 between Guadeloupe and Antigua. About 6,000 persons were killed in Guadeloupe. Other significant earthquakes include the Port Royal (1692) and Kingston (1907) events and the 1842 event in Hispaniola. All four events claimed more than 1,000 lives. The January 12, 2010 earthquake in Haiti resulted in more than 200,000 deaths, 8 billion in losses (CEDMA, 2010). Most earthquake related casualties occurred in the Greater Antilles. These islands are also the most populous, and most of the deaths were due to secondary hazards such as landslides, tsunamis, liquefaction and fire.

The hazardousness of this region is not in question.

TABLE 2 Major hurricanes in the Caribbean by decades, 1970–2008

Category of hurricane	1970–79	1980–89	1990–99	2000–08
Category 3	9	7	11	10+
Category 4	3	7	12	12+
Category 5	3	3	2	7+
Total	15	17	25	29+

2.2. Episodes of change in DRR in CDEMA states

The first period of significant intervention in the area of institutionalized disaster management took place subsequent to 1979 after a number of large impacting events in the region, starting with the 1974 earthquake in Antigua and Barbuda, followed by a series of hurricanes in 1979 (Hurricanes David and Frederick), the volcanic eruption in Vincent and the Grenadines in 1979 and extensive flooding in Jamaica also in 1979, and then in 1980 Hurricane Allen (Collymore, 1987).

These were followed by a decade of diverse impacting events which many communities in the Caribbean region stretching from as far south as the Windward Islands and going as far north as the Greater Antilles, with significant loss of life, substantive damage to property and loss of revenue (Table 3).

Out of these series of events in the 1970s emerged a consortium of concerns about the readiness of the region and especially the states to at least minimize human losses through efficient and coordinated response and relief. Additionally, the events generated a movement for the promotion of the reduction of the substantive socio-economic and especially physical losses associated with impacting hazard events.

TABLE 3 Hazards impacting CARICOM states: 1988–1999

Hazard	Year	Magnitude	Estimated cost	Countries affected
Hurricane Hugo	1989	Category 4/5	$759.4m	Antigua and Barbuda British Virgin Islands Dominica, Montserrat
Tropical Storm Debby	1994	Less than 74 mph		S Lucia
Iris/Marilyn/Luis	1995	Iris – Category 3/4 Marilyn – Category 1 Luis – Category 3	$1,858m	Anguilla, Antigua and Barbuda, Dominica Montserrat, St Kitts-Nevis
Georges	1998	Category 3	$1,195,950m*	Antigua and Barbuda Dominica, St Kitts-Nevis
Lenny	1999	Category 4/5	$712,521,757	Anguilla, Antigua and Barbuda, Dominica Grenada, St Kitts-Nevis S Lucia, St Vincent and Grenadines
Volcano	1995 to present		1995 – Negative growth rate of –7.61% 1996 – Negative growth rate of –20.15%	Montserrat
Landslides	1997	Layout		Dominica
Hurricane Gilbert	1988	Category 4	Estimated losses US$4 billion. Widespread devastation	Jamaica
Hurricane Floyd	1999	Category 4		Bahamas

*Figures for Dominica not included.

In this regard, we saw an initial discussion among key donors and players in the region about a regional project to focus on developing our preparedness capacity as well as putting on the table issues of preparedness and prevention. This later emerged in the form of the multi-donor-supported Pan Caribbean Disaster Preparedness and Prevention Project (PCDPPP). It also represented a first effort at donor cooperation around a common goal of disaster preparedness within a regional context. This was also accompanied by a sense of greater consciousness, at the national level, of the need to put in place an infrastructure for improving public awareness and education on the threats posed to the populations, especially by the hurricane hazard. In this regard, we saw the establishment of the Office of Disaster Preparedness and Emergency Response Coordination in Jamaica and more focus on public education activities in Barbados. These initial interventions were associated with increased public education activities, largely hurricane centred and hurricane season driven.

- The birth of PCDPP was associated with a shift in disaster-related programming from relief distribution to disaster preparedness, especially response. The initial activities of this regional intervention included increasing disaster awareness training of trainers, needs assessment and initial workshops in Member States aimed at sensitizing political and administrative decision makers. A key result of the PCDPP programme was a cadre of sensitized technocrats and decision makers convinced of and committed to the need for organized national activities aimed at increasing disaster preparedness.

The hazard impacting experiences of the 1970s and early 1980s highlighted the inability of the states of the region to effectively coordinate response to natural hazard-induced crises. The agenda here focused on the development of standard notification procedures, early warning systems and emergency telecommunications.

Again, Jamaica and Barbados were the prime sources of experience and expertise in hurricane preparedness plans. The warning alert and notification procedures used by the ODP and the CERO formed the basis of refinement and development in other states.

It was recognized that equally important in the management of an emergency is the ability for key responders and decision makers to communicate with each other. The high priority given to this need can be gleaned from the fact that the PCDPPP agreed to raise the resources to ensure that an HF radio system was installed in key meteorological stations in the region as well as in police headquarters. Thus in spite of the failures of the commercial telecommunications systems, emergency managers were still able to communicate with neighbouring states to receive or give general damage assessment and emergency needs. Hurricane Hugo demonstrated the extent to which this foresight was pragmatic (Caribbean Disaster News, 1989).

Two complementary activities were developed to ensure the efficient use of equipment. A bi-weekly call-up programme was instituted, where all states on the network checked on specific days at designated times on a designated emergency frequency. Valuable experience was gained in message handling under varying propagation conditions.

In addition, an annual disaster simulation was held in a different Member State each year. The simulation tested the capability of the national emergency system to handle an emergency with disaster potential. At the same time, it included other regional personnel and resources, widening the educational experience.

3. Reinforcing the initial programmes

The initial programmes in disaster management in the Caribbean were designed to increase disaster awareness and improve response capabilities. To ensure the continuity of these achievements, attention was directed to developing human resources. Thus some of the project funds were allocated to exposing Member States nationals to short training programmes within and outside the region. Examples include sponsoring telecommunications officers to the USTTI to participate in emergency telecommunications courses.

At the same time, the non-governmental organizational (NGO) resources were recognized and efforts were made to integrate these. Thus, ham radio operations also benefited from the above training. More importantly, they became more organized to play a critical role in both national and region-wide disasters. The formulation of the Lesser Antilles Weather and Emergency Network is the manifestation of this sharing of regional experiences.

4. Moving beyond preparedness

The activities and programmes of disaster management in the region resulted in the significant reduction of loss of life and injuries from disasters affecting the region. However, the losses to both private and public sector assets and resources continued to increase. The need for disaster prevention and mitigation programmes was clear. An agenda for loss reduction was formulated, which sought to incorporate all interested and affected publics. Some of the related activities are briefly summarized below.

4.1. Vulnerability assessment

To eliminate losses, it was important to know what the vulnerable elements were. In light of the severe destruction to designated hurricane shelters in many islands, these became the first targets of vulnerability evaluation. Initially this was done under the auspices of PCDPPP-funded evaluation teams. During the evaluation exercises, the teams collaborated with the local ministry of public works and engineering professional organizations.

The benefit of this approach was that two key resource groups important to the evaluation of critical buildings and infrastructure were exposed to a skill, which now became locally endowed.

A good example of this contagious skill diffusion was evaluation of hurricane shelters in Barbados executed by the Barbados Association of Professional Engineers in 1987.

At the level of the community, some attempt was also made to infuse vulnerability concepts. Local governments, NGOs and private sector leaders in S Lucia were the first beneficiaries of this effort. The process and results were documented and shared in Grenada. In an effort to effectively respond to requests from other Member States for this kind of training, a training video and institutional booklet in community vulnerability assessment have been prepared, and these are in the distribution stage.

4.2. Prevention and mitigation

The disaster management community of the Caribbean recognized that to take disaster management beyond preparedness would entail the strengthening of national emergency organizations and the institutionalizing of the disaster management process. The strategies pursued were therefore multifocal.

Through the national focal points, governments were lobbied and they agreed to allocate resources to disaster management in their recurrent budgets. The amounts and modalities on the allocations varied from state to state. What was important was the recognition of disaster management as one of national concern.

In addition, national disaster coordinators (NDCs) were exposed through various activities to the training necessary to pursue their expanded roles. Between 1982 and 1984, more than 10 such workshops were held in the PCDPPP region and covered all aspects of emergency management.

NDCs were also encouraged and supported in visiting organizations in other Member States.

At a higher level of organizational structuring, considerable bilateral cooperation took place. The Government of Jamaica responded to the Government of Antigua and Barbuda's request for technical assistance in developing an organizational structure for emergency management and a review of its National Disaster Plan. The result is that Antigua and Barbuda now has a Department of Emergency Services and a draft National Disaster Plan in the early 1990s. The result is that Antigua and Barbuda was able to establish a disaster management office and a draft National Disaster Plan.

Trinidad and Tobago also benefited from regional technical expertise in the structuring and establishment of its National Emergency Management Agency. In four years, the twin-island Republic has moved from a situation where no central organization existed to one that is well staffed with links to key resource groups.

4.3. Information networking

Hurricane Gilbert, which struck Jamaica in 1988, highlighted the need for greater information networking. At both the national and regional levels resources related to emergency management were unknown. Consequently, supplies and skills, which were adequately available in the region, were requested from elsewhere. The zinc phenomenon in Jamaica is a fall out of that information gap (Collymore, 1989).

In the aftermath of Hurricane Gilbert, the Government of Jamaica in collaboration with the OAS established an emergency information system, which sought to catalogue all critical national resources, giving location, vulnerability, contract points and accessing protocols. It seeks to utilize the day-to-day information captured systems and provide a mechanism for selective use across national institutions. PCDPPP members have reviewed the system and found it to be very desirable. The national disaster focal points in other states therefore sought to lobby such a resource.

Technical guidance in the development and refinement of the system within the region will come principally from Jamaica, which has had more than two years of intense implementation experience with EIS at the time.

4.4. Incorporating affected publics

The recognition and high priority given to acquiring the widest possible participation in disaster management in the region is attested by the initiatives and results outlined below.

4.4.1. NGO involvement

The approach here was to see NGO resources as complementary to those of the government. It also recognized that NGOs usually have direct and sustained contact with many communities and that any attempt at improving disaster management at the community level must incorporate NGOs.

Consequently, a meeting of regional NGOs convened to review their roles in disaster management and refine strategies for collaboration and cooperation in their execution. The result of the 1989 meeting has resulted in a heightened involvement of regional NGOs in disaster management. The process of role delineation in the various phases of disaster management continues to be refined.

4.4.2. Professional groups and voluntary organizations

The extent to which the process of consultation, integration and involvement has paid dividends is provided by the experience with the Caribbean Council of Engineering Organizations (CCEO).

Against the background of persistent and extensive damage to buildings and other critical facilities resulting from hurricane impacts, there was a cry for minimum design and building standards. The cry was given leadership and structure by CCEO, whose members were part of the initial target groups sensitized by the PCDPP awareness programmes with the alliance of CARICOM and the support of international donors, especially the CDB. The Caribbean Uniform Building Code has been developed and is now being promoted.

Other professional groups such as architects and planners are also beginning to lend their informed perspective to the development and refinement of land-use-related policies that can facilitate disaster loss reduction.

4.4.3. Media cooperation

Over the years, the media has been accused of being concerned only with the sensation of disaster impacts. In increasing the disaster awareness of media professionals. In the aftermath of Hurricanes Gilbert (1988) and Hugo (1989) when several electronic media outfits were incapacitated, there has been an overwhelming response in increasing the disaster awareness by media professionals.

After a primary meeting of regional media managers facilitated by the PCDPPP, several definite actions have been undertaken by the media itself. Among these are the following:

- Preventive procedures to ensure survival of the radio broadcast service.
- Strategies for providing information to affected communities, which have lost services of the commercial radio.
- Cooperation on resource and information-sharing in constraint circumstances.
- It is important to note that in all cases, efforts were made to work through established national and regional activities.

The second major episode in institutional arrangements for disaster management occurred about 10 years later following Hurricane Gilbert in 1988 and Hurricane Hugo in 1989. The multi-island devastation associated with the Hugo event in the Eastern Caribbean islands, following the earlier Hurricane Gilbert devastation in Jamaica, consolidated in the minds of our political leadership the need for a structured mechanism to allow for the management of its collective interest in being responsive to the humanitarian needs of its community.

This resulted in the establishment of the Caribbean Disaster Emergency Response Agency in 1991. It represented a new dimension in the disaster management arrangements focused on intergovernmental cooperation as well as better structuring of inter-agency cooperation in support of disaster preparedness and response initiatives. It provided a framework for supporting the capacity enhancement of national focal points and introduced initial elements of inter-agency programming consultations as a premise for better rationalizing resources and efforts towards improving capacity in the area of disaster management in this region. During this decade, associated with the International Decade for Disaster Risk Reduction, there were several additional drivers that gave momentum to these efforts to structure national platforms that sought to propel disaster management beyond response and relief (Figure 1).

5. Post-PCDPPP

The third episode is associated with the closure of PCDPP in 1991 and the establishment of the Caribbean Disaster Emergency Response Agency (CDERA) and galvanized by a series of disaster impacts centred around the events of 1995: Hurricane Luis/Marilyn/Iris, the Montserrat Volcanic Emergency and later Hurricanes Georges 1998, Floyd and Lenny 1999. Coming within a very short period they raised significant concerns about repeated losses to the economic gains of struggling developing economies as well as issues about the sustainability of our development in the context of this exposed vulnerability.

This period, which is still in an evolutionary phase, has seen a different and significant departure from the approach to disaster management previously pursued. Even though the PCDPP

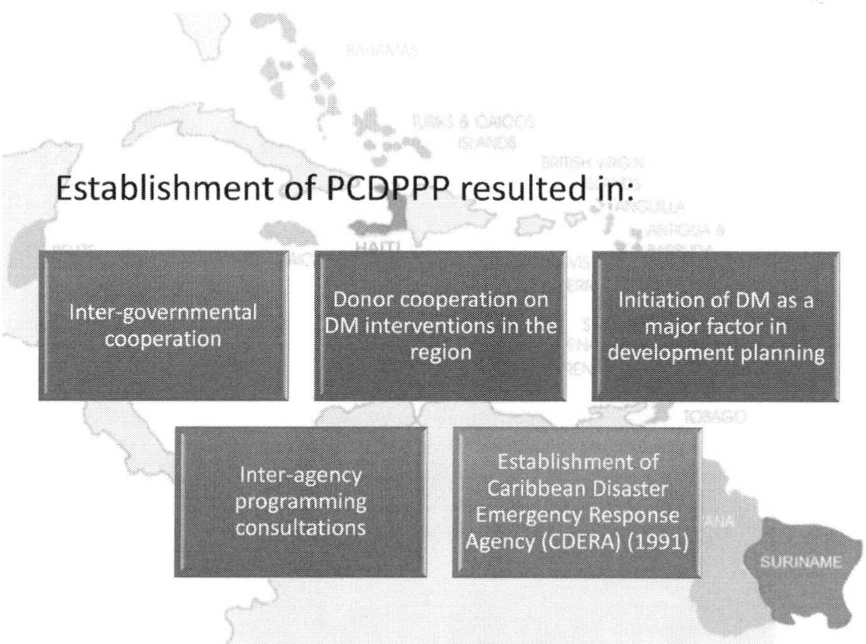

FIGURE 1 Drivers of change in disaster management approach

project sought to look at issues beyond preparedness, the main conditions existing in countries, with little or no capacity for managing disaster events, led to a necessary heavy focus on basic infrastructure for preparedness, including shelter-related vulnerability assessments. The 1990–2000 epoch reaffirmed the importance of being more strategic and holistic in efforts to reduce the extensive loss of property and socio-economic gains. First of all it recognized a strong link between disaster management and development, sought a more committed effort in terms of resources and basic infrastructure for supporting national disaster organizations (NDOs), and also to place the issue of risk management in the broader context of government decision making especially in the productive sectors. It therefore broadened the dialogue from disaster offices and organizations to economic sectors and institutions and saw the rapid involvement of multi lateral and bilateral organizations and multi lateral financial institutions in this process (see Figure 2).

The event-driven developments in disaster risk reduction (DRR) programming and embryonic policy suggest a high measure of reactiveness. The question that must therefore be asked is whether or how is the region seeking to be more proactive in addressing DRR issues.

An examination of current approaches to disaster management suggests that there is some momentum towards the disaster management programming agenda that is proactive, broad in focus and reflects the diversity of threats we face. While we have had significant investments in the basic institutional infrastructure, that is, increased staff, facilities and some limited budget, we still have some fundamental gaps to address in disaster management (Jones, 2006; Collymore, 2005, 2008).

Disasters serve as evidence of the need for change in public policy and practice and create opportunities to redesign, revise or rebuild the human environment damaged by the event (Comfort et al., 1988).

The Comprehensive Disaster Management Strategy Framework, developed in 2001 and revised in 2007, represents the first step in this direction. Comprehensive disaster management is a concept that seeks to incorporate management of all hazards, at all phases of the continuum, by all segments of civil society, inclusive

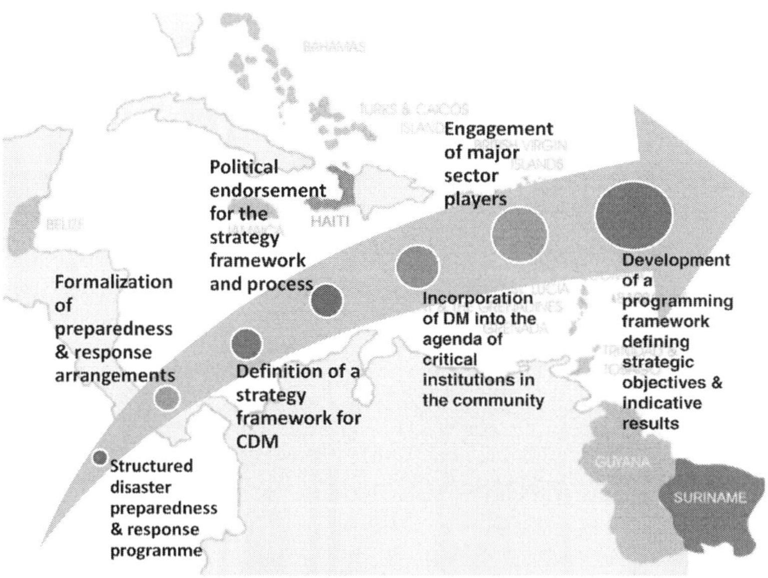

FIGURE 2 Disaster management approach since PCDPPP

of the public and private sectors, non-governmental and service organizations, urban and rural communities, and the general population living in disaster-prone areas. It is this recognition of the inextricable link between disasters, development and the environment that informed a consultative dialogue among key regional stakeholders to fashion a framework for structuring this integration (CDERA, 2001). This has been further enhanced by building on lessons learnt in the initial five-year period (CDERA, 2006; UNDP, 2007).

The goal of the Enhanced Regional Strategy and Results Framework is sustainable development enhanced through comprehensive disaster management. Its purpose is to strengthen regional-, national- and community-level capacity for mitigation, management and coordinated response to natural and technological hazards and the effects of climate change (as seen in Figure 3).

It is anchored around four priority outcomes articulated in results and programme base language:

(i) Enhanced institutional support for CDM programme implementation at national and regional levels.
(ii) An effective mechanism and programme for management of CDM knowledge.
(iii) Disaster risk reduction mainstreamed at national levels and incorporated into key sectors of national economies (including tourism, health, agriculture and education).
(iv) Enhanced resilience in CDERA states/territories to mitigate and respond to the adverse affects of climate change and disasters.

The associated programme of work is aligned to the Hyogo Framework for Action (HFA) adopted by the 2005 United Nations World Conference on Disaster Reduction. The strategic framework is a platform from which disaster loss reduction interventions are being structured, cooperative programming initiated and partnership management elaborated.

In seeking to put integrated comprehensive disaster management (CDM) into operation at

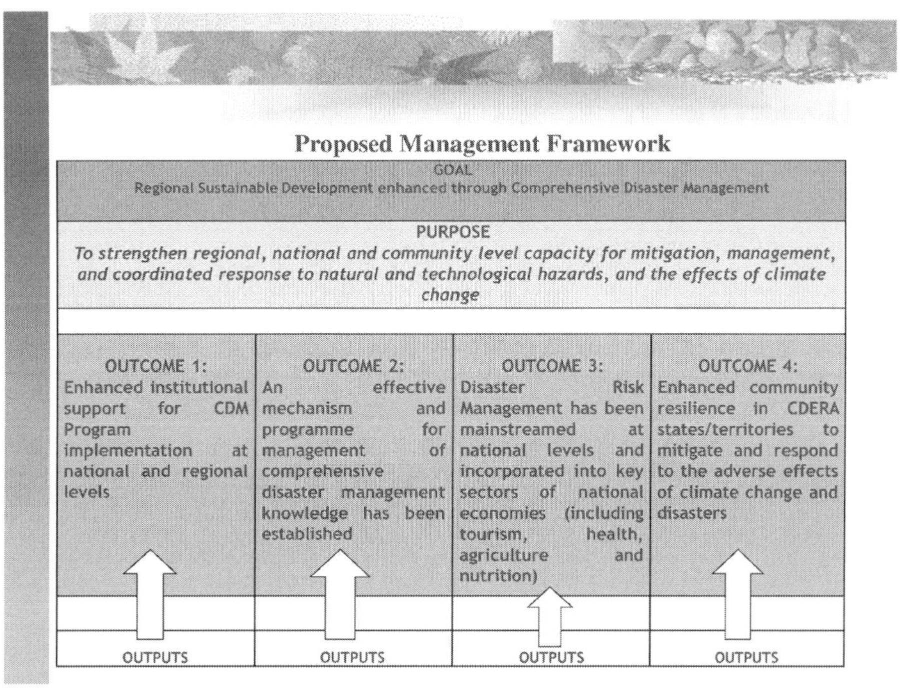

FIGURE 3 Comprehensive disaster management

the national and regional levels, disaster management organizations supported by their respective governments are being required to assess their institutional capacities to enable them to operate within the expanded mandates and make provision for the appropriate levels of human, financial and material resources, if they are to achieve acceptable levels of public safety and support sustainable development goals.

This is a clear mission of the CDM partnership and is reflected in the focus and priorities of partner support for DRR in the CDEMA states. Priority Outcome 1, capacity of national and regional organizations to implement CDM, and Outcome 4, capacity for community resilience strengthened, have been the primary beneficiaries of technical support. More than 60 per cent of the resources for 2007–2012 of the more than estimated US$100m associated with programme of work of December 2009 have directed towards these two key priority outcomes (CDEMA, 2009).

What is important about the strategy is that it has been developed as part of a consensus consultation among private sector, public sector and civil society, key stakeholders have been involved in its design and are engaging the ownership of this and translating the principles of this strategy into the agenda and programme of its constituents.

The following are indicative of the commitment to this agenda and process.

5.1. Supporting national CDM implementation

This embraces a diversity of initiatives designed to promote political awareness and buy-in, promotion of the identification and support of national CDM champions, assistance in the design and monitoring of multi-year result-based risk reduction programmes, DRR baselining and knowledge enhancement. Much of this has been supported by bilateral partners with major financial inputs from the EC, CIDA, DFID, JICA and Government of Austria. The involvement of regional institutions of the Caribbean Community (CARICOM) has also been critical in helping to deliver the national-level technical assistance and so too has been the support of the multilaterals and Multi-lateral Financial Institutions.

5.2. Strengthening CDM cooperation

The initiatives here have focused on establishing a facility for the capture and analysis of partner interventions in DRR in the CEDMA states, partner programming consultations, organization of thematic clusters, a regional networking and experience sharing forum (CDM Conference), and a reporting and governance mechanism. This is now wrapped within the framework of a blueprint for CDM (DRR) at the national level that seeks to document and promote agreed standards and models for national DRR policy, associated legislation, organizational structure and performance benchmarks. The many issues and processes associated with CDM implementation are outlined in Figures 4–7.

CDM has become a Caribbean HFA brand. It can be utilized to further promote commitment to the DDR agenda through the process designation of CDM-recognized organizations, partners and CDM-friendly businesses.

6. Beyond the CDM

In this regard there will be a need to look at practices, mechanisms, products, tools and services that serve to promote a culture of loss reduction, whether this is centred on the preparedness side, response side or recovery side. This client-centred mainstreaming of CDM represents the next critical frontier to accommodate a proactive approach to managing the change necessary to sustain the emerging interests in multi-hazard, multi-sector and multi-phased disaster management interventions.

This proactive approach must be informed by lessons of our past and current experiences. It will inevitably require some serious reflection on what we mean by institutionalization and the implications of this for the nature and framing of DRR policy.

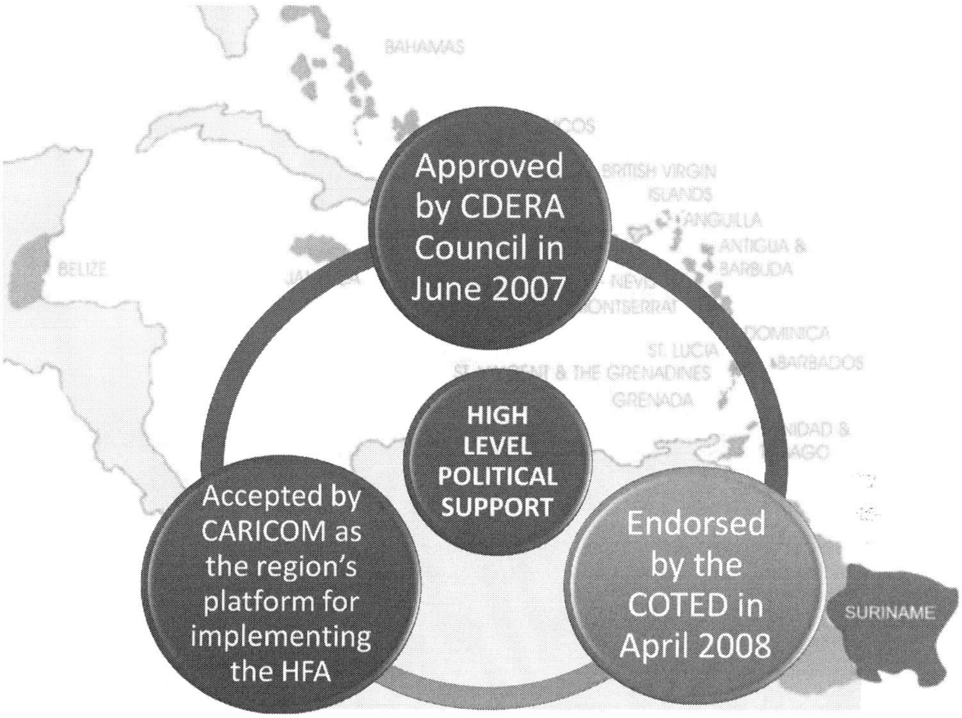

FIGURE 4 High-level political support

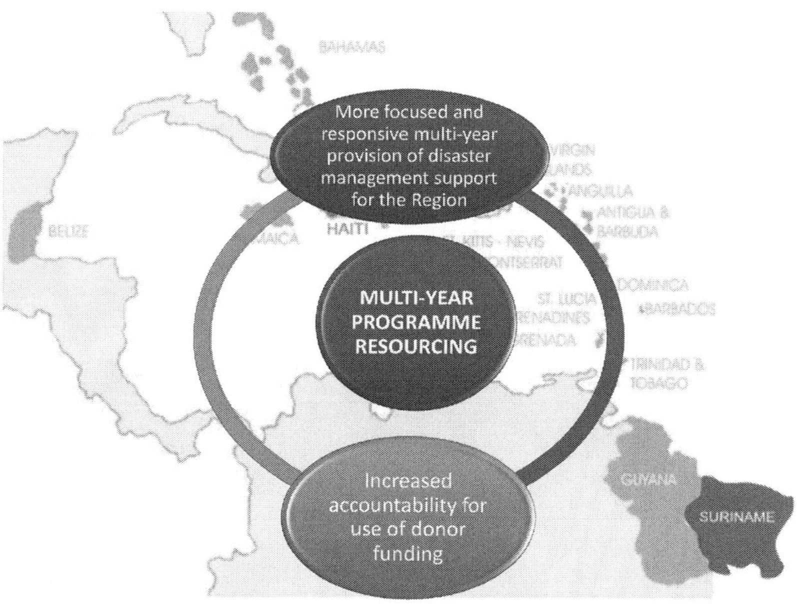

FIGURE 5 Multi-year programme resource

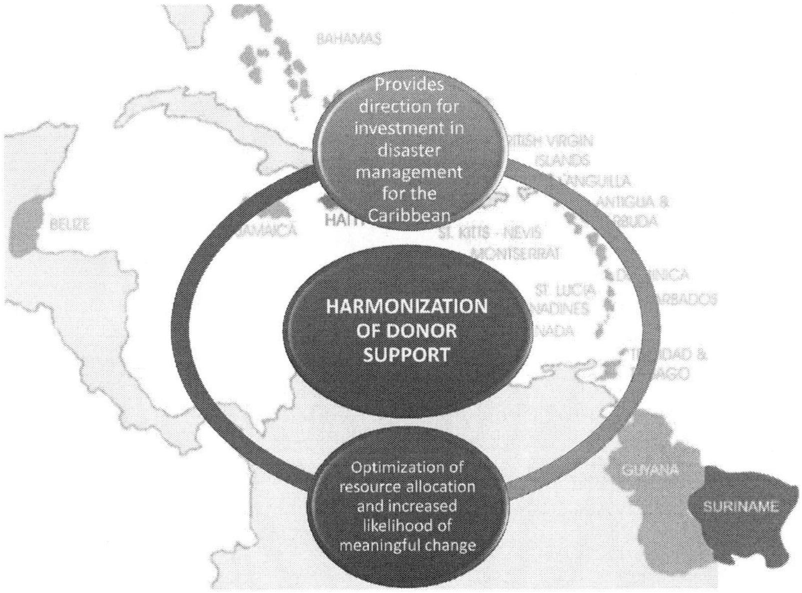

FIGURE 6 Harmonization of donar support

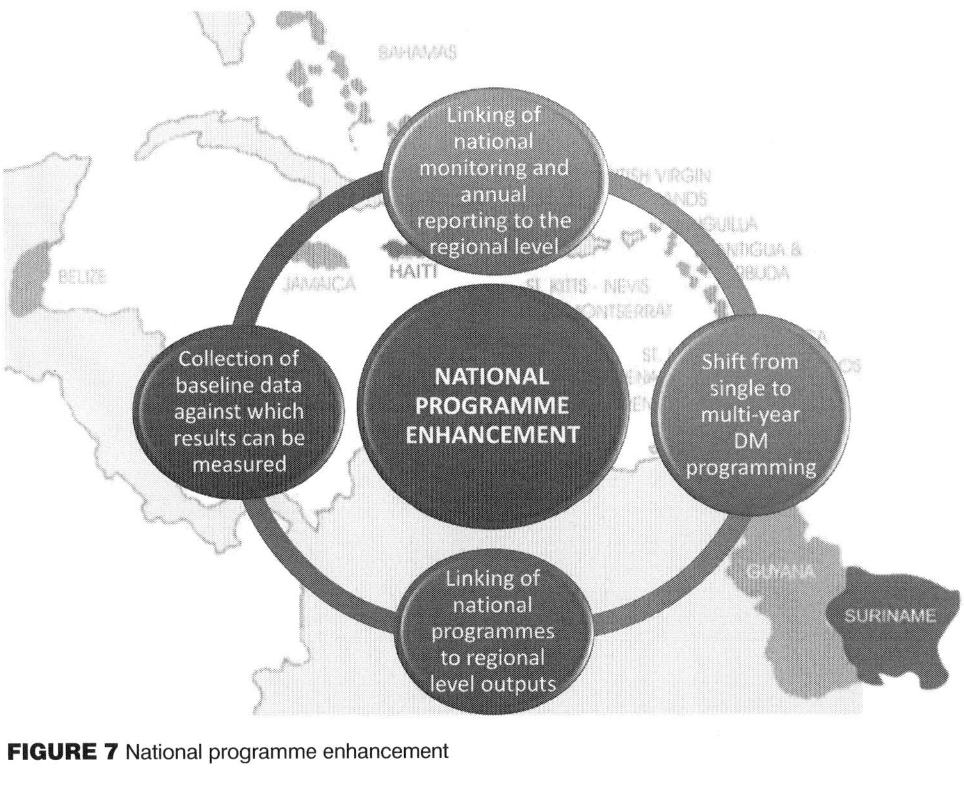

FIGURE 7 National programme enhancement

The CDM strategy has suggested that, to the extent that national strategy and regional frameworks exist, DRR implementation priorities are more likely to be beneficiary centred and less likely to be donor driven. It also facilitates a broader lens for looking at risks to be managed beyond those driven by seasonality and impact experience. It will better allow for addressing the observed disconnection between disaster management and development. In this regard we will need to ask ourselves some fundamental questions about disaster management at the level of the state.

The first question has to do with what levels of risk states are prepared to accept. This will inform the balance of interventions in preparedness and mitigation and how these relate to other development goals. It will also require that significant attention be paid to the definition of who are the vulnerable and how the vulnerability is created. Indeed I wish to posit that this contextualization of vulnerability is critical for interdisciplinary and cross-ministry programming linkages infrastructure necessary for linking DRR to the development agenda.

How do we propose to structure the solutions? Are these going to be technology driven? Are we going to examine social resilience and seek to strengthen this? These questions speak to the issue of who will be the players and what will be the process through which we will seek to develop the policies and structure the interventions. What assumptions are we making about vulnerabilities? Is it the disposition that the vulnerability we have, or we have identified, is a reflection of a lack of policy support articulation and resource commitment or will it (the vulnerability) be unchanged in spite of change of policy or resource commitment? The answers to these questions will raise concerns about the relevance and appropriateness of what is being proposed.

What may now be necessary is an examination of how the involvement of key players in the planning and decision-making corridors of the states are engaged in the CMD exercise.

As we seek to build this culture of DRR, an expected impact of CDM, much more thought will have to be given to a shared understanding and articulation of 'institutionalization'. The scope will need to go beyond organizational strengthening to the considering elements of social behaviour and culture and seek to impact them in such a way that issues of risk reduction and safety become one of our accepted mores.

Similar dialogue will be required on the articulation of capacity development and/or strengthening. There will be a need to ask what is the capacity to be built for; is it the capacity to respond, for better decision making or for more informed decision making? Do we want the capacity to be in a centralized mechanism, be a decentralized community-based capacity or a combination of the above? What are the institutional mixes that would allow us to achieve these?

The above questions have not been addressed in an open and explicit way because our disaster management programmes have largely been reactive to experiences in our islands despite the repeated history of development dislocation associated with impacts.

The emerging agenda for CDM will also have to look at the process elements in institutional capacity building and pay more attention to issues of skills and technology transfer, the nature of public education and training, public information, its structure and delivery modes.

There is an urgent need to re-assess our internal capacities to deliver on some of these issues and to see this as the first line of intervention in our programme development. There is too much emphasis on external support as a first-line intervention of disaster management activities, and therefore our programmes have become very dependent on external interventions. This creates significant problems for sustainability and it undermines self-reliance initiatives.

A critical link to all of this is a national consciousness of economic readiness for managing these threats, their impacts and consequences. While we have consistently focused on individual, family and business planning, the State has not addressed in an integrated and coordinated way the economic preparedness policies and programmes that it needs to put in place to ensure that it can survive and recover in the shortest possible time to service its constituency of public services, private sector essential requirements and the civil society supporting services in a crisis-related environment. This is where we believe that the engagement of the private sector and the civil society in a structured format will allow for a broad-based inducement of our political leadership to provide the higher level of strategic support for building the culture of DRR.

Elsewhere I have addressed the issue of identifying opportunities for accommodating some of the changes and approaches referred to above. Here I will simply refer to the links with the coastal zone management programmes, GIS capacity enhancement as a platform for common information sharing, consolidation of risk and development opportunities and investment options, poverty reduction and alleviation to reinforce the principles of disaster management as a cross-cutting entity in our sustainability development portfolio and in particular the issue of climate change, which thus far has only marginally engaged the disaster management entities in addressing these matters. As I have argued elsewhere, the experience in managing current climate variability, whether this is in agriculture, building design or fishing systems, is important if our consideration of extremities and the adaptation to these extremities is to be informed by an appreciation of our current resilience. To date there has only been a token engagement of disaster management communities in this process, and therefore we welcome the adoption of the Caribbean Community Strategy for Climate Change and Development. The need for this interfacing is recognized in the CDM Strategy goal, and already efforts are being made to bridge the CCA and DRR programming.

7. What does all of this mean for National Disaster Offices (NDOs)?

First of all, if NDOs are going to play a meaningful role in the CDM framework, they must have a legal basis for their existence. This is still to be formalized in some states. NDOs must be given the capacity to provide coordination support and strategic direction and to really be the broker for the national advocacy efforts. Leadership capability now needs to become a key requirement for leading the NDO. CDEMA, with the support of the government of Australia, is already seeking to design and test a leadership module in the DRR learning, with the goal of having this regularized and offered on a regular schedule to the States.

A major issue that will emerge in this transformation is the role of the traditional disaster office. The up-scaling of the scope and target of DRR and existing concerns about the capacity of NDOs, even of CDEMA Coordinating Unit, make this a necessary and important debate that must be encouraged. There is a need to demystify ourselves of the notion that the disaster management office needs to undertake all the critical interventions in risk management. Most of the existing risks are in key production sectors and enterprise zones, and therefore we must be working to engage those parties as part of this process. The institutional arrangements for engaging the disaster office as an advisory and advocate forum need to be considered and its expectations managed so that the lack of action in risk management does not affect the credibility of the advocacy body itself.

Linked to this very critical notion of responsibility must be the erosion of the current culture in public service that disaster preparedness activities are extra impositions. There must be a clear understanding that the issue of public safety and security of public assets is the responsibility of civil service. Current legislation has set in place the foundation for moving in that direction. If only modestly successful, it can be a concrete step in making DRR everybody's business. So too would be the convening of annual parliamentary debate on the status of risk management in each state.

The ideas I have thrown out here are not earth shattering. They may already be shared but not discussed. They also allow one to pose the question as to what ought to be the interventions of the academic and tertiary institutional communities in seeking to address some of the concerns I have raised.

First of all, we need to set a definition of the risk, where it is, how it is constructed in terms of what it says about our social systems and services, and the policies to manage these. We need better social investments and social tools for looking at and influencing community behaviour. We need to look at the systems and economic practices that reduce the resilience of our traditional agriculture and other farming systems. We need to look at organizational management and corporate responsibility as it relates to risk management. There are a plethora of interventions that are possible in this area but have not been detailed.

References

Caribbean Disaster Emergency Management Agency (CDEMA), 2010. *Summary of CARICOM Support to Haiti in the Aftermath of the January 12, 2010, Earthquake*. CDEMA, Bridgetown, Barbados.

Caribbean Disaster Emergency Response Agency, 2004. The Comprehensive Disaster Management Online Database: Online Analyst. Barbados.

Caribbean Disaster Emergency Response Agency (CDERA), 2008. *Review of 2007/2008 Emergency Events in CDERA Participating States*, paper presented to the Eighteenth Meeting of the Board of Directors of the CDERA, St Johns, Antigua, 22–23 May.

CDERA, 2001. Comprehensive Disaster Management in the Caribbean: Baseline Study. Submitted by Jones, E., Bisek, P. A. and Dinstein, C. to CDERA under CDERA/US Agency for International Development (USAID/UNDP) Comprehensive Approach for Disaster Management in the Caribbean Project.

CDERA, 2001a. A Strategy and Results Framework for Comprehensive Disaster Management in the Caribbean. 2001 Submitted by Jones, E., Bisek, P. A. and Conrad Ornstein to Caribbean Disaster Emergency Response Agency (CDERA) under the CDERA/ United States Agency for International Development (USAID)/United Nations Development Programme (UNDP) Comprehensive Approach for Disaster Management in the Caribbean Project.

Caribbean Disaster Emergency Reponse Agency, 2001. *A Strategy & Results Framework for Comprehensive Disaster Management in the Caribbean*; Prepared by Bisek et al. with the Support of USAID and UNDP. Barbados.

Caribbean Disaster News, 1989. 'Telecommunications: The Experience of Hurricane Gilbert'. Issue 18, 1–2 June, UNDRO/Pan Caribbean Disaster Preparedness and Prevention Project. Antigua.

Centre for Research on the Epidemiology of Disasters (CRED). *EM-DAT the OFDA/CRED International Disaster Database Brussels*. www.emdat.be. Accessed 11 March 2010.

Collymore, J. McA., 2008. Beyond humanitarianism: building resilient communities. Revisiting the development dialogue, *Enduring Hazards in the Caribbean: Moving from Reactive to Proactive*, B. Seran (ed.). UWI Press.

Collymore, J. McA., 2005. 2004-2005 Disaster Events Opportunities for Advancing Disaster Management. Paper prepared for the 18th Annual Technical Conference of the Association of Professional Engineers of Trinidad and Tobago (APETT) on Engineering Infrastructure, Planning and Development for Disaster Preparedness and Mitigation, The University of the West Indies, St Augustine, Trinidad and Tobago, 14–15 April 2005.

Collymore, J. McA., 2004. Disaster impacts on the Caribbean. In *International Perspectives on Natural Disasters: Occurrences, Mitigation and Consequences*, edited by Joseph Stoltman, John Lidstone and Lisa M. DeChano. 303–323. Dordrecht, The Nethederlands: Kluwer Academic Publication.

Collymore, J. McA., 1992. Hurricane Hugo – A multi island disaster: further lessons for the Caribbean. *Disaster Management*, 4(3). 163–167.

Collymore, J. McA., 1992a. Planning to reduce the socio-economic impacts of natural hazards on Caribbean society. *The Journal of Geological Society of Jamaica.*

Collymore, J. McA., 1989. Disaster Planning Lessons for the Caribbean. The Gilbert Experience. *Disaster Management*, 2(2). 87–93.

Collymore, J. McA., 1988. Geographical Events and Human-use Systems: A Revised Research Agenda in Nkemderin, L. C. ed. 1988. "The Tropical Environment". *Proceedings of the International Symposium, on the Physical and Human Resources of the Tropics*. Kingston, Jamaica, 4–7 August 1987. 85–91.

Collymore, J. McA., Framing a stake in the global disaster loss reduction agenda: revisiting the development dialogue. Barker, et al. (ed). in press

Comfort, L. K. (ed), 1988. Managing disasters: strategies and policy perspectives. Durham: Duke Press Policy Studies. Reframing disaster policy: the global evolution of vulnerable communities. *Environmental Hazards*. 39–44.

Cummins, J. D. and Mahul, O., 2010. *Catastrophic Risk Financing in Developing Countries. Principles for Public Intervention.* World Bank.

DFID, 2005. *Disaster Risk Reduction: A Development Concern: A Scoping Study on the Links between Disaster Risk Reduction, Poverty and Development.*

Hewitt, K., 1983. *Interpretations of Calamity.* Allen & Unwin, Winchester, MA, USA.

IDB, 2000. *Facing the Challenges of Natural Disasters in the Latin America and the Caribbean. An IDB Plan of Action.* Inter-American Development Bank Environmental Division, Washington, DC.

IDB, 2000a. *A Matter of Development: How to Reduce Vulnerability in the Face of Disasters.* IDB, Washington, DC.

IDB, IMF, OAS, World Bank, 2005. *The Economic of Disaster Mitigation in the Caribbean: Quantifying the Benefits and Costs of Natural Hazard Losses; Lessons Learnt from 2004 Hurricane Seasons*, Working Paper. Washington, DC.

Pelling, and Uitto, J. I., 2002. Small island developing states: Natural Disaster Vulnerability and Global Change. *Environmental Hazards*, 3, 49–62.

Rasmussen, T., 2004. *Macroeconomic Implications of Natural Disasters in the Caribbean.* IMF Working Paper.

UNDP, 2007. *British Virgin Islands, Case Study: Integration of Disaster Risk Management into the Development Agenda. Best Practices in Risk Reduction.* Prepared for the Caribbean Risk Management Initiative by ESL Management Solutions.

UNDRO, 1991. *Mitigating Natural Disasters. Phenomena Effects and Options.* United Nations, New York.

UN OCHA, 2007. *Record Number of UN Emergency Missions in the Americas.* 26 December. www.reliefweb.int/rw/RWB.NSF/db900SID/SHES7A9MHU?OpenDocument. Accessed 11 March 2010.

UN-OHRLLS/AKC, 2008. *The Impacts of Climate Change on Least Developing Countries and Small Island Developing State.*

ENVIRONMENTAL HAZARDS
Human and Policy Dimensions

The Caribbean Disaster Mitigation Project

KEITH FORD*

7 Oldgate Drive, Kingston 19, Jamaica, West Indies

This paper highlights the development of the United States Agency for International Development's (USAID's) Caribbean Disaster Mitigation Project (CDMP). The CDMP was developed in 1992 by the USAID Regional and Urban Housing Office and the Office of Foreign Disaster Assistance (OFDA). The development of the CDMP was influenced by several international and regional events. Primary among these were the declaration of the International Decade Disaster Reduction and the development of a global Prevention, Mitigation and Preparedness Strategy by the US OFDA. This paper examines the impact the International Disaster Mitigation conference of 1984 and the 1985 'Pan Caribbean Workshop on Design Mechanisms for Reducing Vulnerability to Natural Disasters' held in Trinidad had on its development.

Keywords: development; disaster mitigation; joint disaster mitigation assessment; regional

Preface

> A great deal of policy thought and policy attention is directed towards managing and mitigating international trade vulnerability and financial vulnerability. It is high time that serious effort be also directed towards managing and mitigating natural hazard vulnerability. Progress in reducing and better managing natural hazard vulnerability will make progress in managing and mitigating international trade and financial vulnerability more sustainable. They are complementary objectives, as nature has a habit of reminding us. It is high time to treat them as such (Compton Bourne, Opening Remarks at the 33rd Annual Meeting of the Board of Governors of the Caribbean Development Bank).

This paper looks at the early beginning of the Caribbean Disaster Mitigation Project, specifically the forces that shaped its development and the general thinking behind its conceptualization.

1. Introduction

The CARICOM Community covers an area of over 3,000 square miles encompassing 16 independent countries and several British, French and Dutch dependent territories. In addition to the four major language groups – English, Spanish, French and Dutch – several dialects are also spoken. With the exception of Belize, Guyana, Suriname, Turks and Caicos, the Bahamas and Bermuda, the countries are located in or around the Caribbean Sea and on the Caribbean Plate.

The region is divided into three sub-groupings: the Greater Antilles, which is composed of the geographically larger islands, the Lesser Antilles consisting of the Leeward and Windward Islands, and the countries located on the Central and South America mainland.

The region is highly vulnerable to both natural and human-made disasters. Data indicate that over the period 1846–1978, 34 per cent of all recorded disasters occurred in the Latin America–Caribbean region, and caused over one million

■ ***Email: fordkb@hotmail.com**

doi:10.3763/ehaz.2011.0003 ISSN: 1747-7891 (print), 1878-0059 (online) www.earthscan.co.uk/journals/ehaz

deaths (Jovel, 1982). Seasonal events such as hurricanes and floods impact the region on a regular basis, while other hazards such as earthquakes and volcanic eruptions pose a constant threat to most of the region. Over the last 20 years, there has been a gradual increase in the threat of technological disasters, particularly in the area of oil spills. This is most noticeable in Trinidad and Tobago, where the number of spills increased from 89 in 1984 to 245 in 1997. The region also recorded 13 chemical spills between 1984 and 1996 (International Federation of Red Cross and Red Crescent Societies, 1997).

Several major disasters stand out in the history of the region. Primary among these are the Port Royal earthquake of 1692 that destroyed the town of Port Royal, with a section of it sinking beneath the sea and drowning hundreds of people (see www.earthquake.usgs.gov/earthquakes/world/events/1692_06_07.php).

The great hurricane of 1780 struck the islands of Barbados, Martinique and Sint Eustatius in the Lesser Antilles over the period 10–16 October 1780. This event resulted in the death of over 22,000 people onshore and several thousands offshore as it devastated the British and French naval fleets (see www.sciencedaily.com/articles/g/great_-hurricane_of_1780.htm).

The eruption of Mount Pelee in Martinique on 8 May 1902 destroyed the town of St Pierre, as hot gases enveloped the town resulting in over 28,000 deaths (see www.geology.sdsu.edu/how_volcanoes_work/Pelee.html).

These events, however, pale in comparison to the devastating earthquake that affected Haiti in January 2010, which resulted in the death of over 250,000 people and totally destroyed the city of Port-au-Prince.

While cognisant of the overall vulnerability of the region, the Caribbean Disaster Mitigation Project (CDMP) was developed primarily to support independent English-speaking countries receiving support from the United States Agency for International Development (USAID). These included Jamaica, Barbados, Belize, Guyana and English-speaking countries in the Leeward and Windward island chain.

2. Factors impacting the development of the CDMP

2.1. The beginning

In 1992, USAID, Regional Housing and Urban Development Office for the Caribbean (RHUDO/CAR) and the Office of Foreign Disaster Assistance (OFDA) were mandated to develop a regional disaster management project that would be consistent with the established USAID Latin America and Caribbean Policy, stated as:

> Improved environmental and natural resources management will contribute to the lessening of the impact of natural disasters and reduce the extent of disaster response needs. Catastrophic and structural risk assessments, code modification, involvement of private insurers in prevention, and community readiness can reduce the damage and hardships when disasters occur. Through a regional activity USAID will work closely with Caribbean governments and the private sector to better prepare for and mitigate the effects of disasters.

The mandate from USAID/LAC was preceded by a joint disaster mitigation assessment exercise conducted in 1991 by RHUDO/CAR and OFDA. The purpose of this exercise was to assess the training needs of disaster managers in the region, their response system, staffing levels and long-term plans. Based on the findings of this assessment, RHUDO/CAR under a joint agreement would develop country-specific projects to address identified needs.

The results of the assessment revealed consistent similarities between countries, with key issues such as weak building inspection procedures including a lack of enforcement of building standards, inappropriate land use, unavailability of hazard-specific maps and the small staff complement of disaster offices being the most cited.

Because of these existing similarities, it was possible to limit the range of proposed activities and instead focus on those considered the most urgent by the responding countries.

Having then identified the priorities, the approach for undertaking these activities needed to be developed and became a major objective of the CDMP. A breakthrough came during the conference workshops of 1984 and 1985 that subsequently provided the inspiration for moving forward with development of the project.

2.2. Conceptualization

Based on the established policy and mandate, it was determined that the project to be designed should be regional in scope with a practical orientation. Specifically, the project was directed to focus on the development and application of practical tools as opposed to 'studies'. Where studies were to be carried out, these were expected to lead a practical output, which the project would be able to implement during the projected time frame.

A major finding during the assessment was the lack of integration of risk reduction measures in the development planning process. This was primarily due to the significant information gap that existed in terms of the availability of scientific data such as hazard and vulnerability maps, combined with the low involvement of the private sector and civil society, in both disaster management and development planning. This lack of inclusion severely restricted the capacity of disaster managers to actively intervene in the development planning process. Consequently, development planners were also unable to actively integrate disaster risk reduction activities in development plans, as a result of which decisions regarding the location of residential and industrial development were often made without the empirical data to justify the decision to locate within a particular area. The inability of planners to effectively provide the technical data necessary to guide the development process led to a situation in which private developers became the de facto planners. In this hazy environment, development projects were often planned and implemented based on political expediency rather than on a thorough assessment of the proposed development by planners.

A significant factor in the formulation of the CDMP was its potential to fill the data gap and to provide an effective platform for engaging the three sectors in disaster management.

Based on the result of the assessment, the overall goal of the project was therefore determined as being: 'To establish a sustainable private/public sector mechanism for disaster mitigation (in the region) to measurably lessen the loss of life and physical and economic damage and reduce the length of recovery (time)'. The CDMP was therefore designed 'to create a more disaster resistant environment by working primarily with private/public sector institutions to develop and implement disaster Mitigation projects' (PADCO, Inc. Caribbean Disaster Mitigation Project Final Report, 1992).

The identified goal would be achieved through the implementation of a number of interrelated activities, primarily:

- Risk audits for electrical, water/sewer systems and key lifeline facilities such as hospitals.
- Hazard mapping to support improved planning and location of physical development.
- The determination of risk insurance based on designs and location.
- Assistance to improve underwriting standards to link the pricing of insurance to actual catastrophic risks.
- The training of builders, artisans, etc. in improved building practices, technology systems and the development of simple guides related to safe building practices.
- The development of strong community projects.

These goals were all identified as having the potential to:

- Reduce the vulnerability of basic infrastructure and critical facilities.
- Improve building standards and practices to reduce natural hazard vulnerability.
- Increase availability and access to natural hazard/disaster risk information for use by

public and private sector developers, investors and insurers.

- Increase community awareness of and involvement in disaster preparedness and mitigation measures.
- Improve availability of insurance and reinsurance for natural hazard perils.

In general, planned activities were seen as increasing the capacity of the host societies to withstand the adverse impacts of natural hazards.

2.3. Influences on the development of the CDMP

The period 1979–1989 was an active one for the region, with several hurricanes impacting the region. The two most damaging storms during this period inflicted damages estimated at US$2.5 billion dollars. Of this amount, US$1 billion was caused by Hurricane Frederick in the Dominican Republic in 1979. Hurricane Gilbert, which struck Jamaica in 1988, inflicted damages estimated at a similar amount. Hurricane David, which struck the Dominican Republic also inflicted significant economic damages there.

The impact of these events on these small island states served as a reference point for the development of the CDMP, which was designed to implement measures aimed at reducing the damage impacts from these and other hazard events. In this regard, both the 1984 conference and the 1985 Trinidad and Tobago workshop played an important role in shaping the final CDMP project document, as both had focused on the integration of practical measures in the development process and the widening of the disaster management base to include the private sector and community groups.

2.4. The 1984 Ocho Rios Conference

The 1984 Mitigation Implementation Conference held in Ocho Rios Jamaica brought together an international gathering of disaster managers to discuss current issues in disaster management. The emphasis of the workshop was primarily on the still developing concept of integrating disaster risk reduction in the development process. Participants recognized the need for hard data such as hazard maps to support this integration, the use of insurance as a regulatory mechanism, and the importance of having a strong and effective building inspectorate. These were seen as being essential in ensuring the integrity of the development process and for improving the resilience of countries.

2.5. The 1985 Trinidad Workshop

The Trinidad Workshop of 1985 was sponsored by the Pan Caribbean Disaster Preparedness and Prevention Project along with several regional and local organizations, including the Association of Professional Engineers of Trinidad and Tobago, the Faculty of Engineering, University of the West Indies, the Council of Caribbean Engineering Organizations, the Trinidad and Tobago Institute of Architects and the Ministry of National Security. The workshop had as its theme 'Design Mechanism for Reducing Vulnerability to Natural Disasters'. This event brought together regional disaster managers, engineers, architects, geologists, geographers, insurance risk adjusters and town planners for them to 'discuss the main aspects of vulnerability to natural hazards and to make recommendations to the regional governments and Agencies on practical methods by which vulnerability can be reduced' (Pan Caribbean Disaster Preparedness and Prevention Project, 1985).

The workshop was organized around the following themes:

- the informal housing sector
- schools
- maintenance
- insurance
- regulations
- utilities
- coastal structures
- flood control
- earthquakes
- planning and land/environment.

Detailed discussions on the above themes by participants with reference often being drawn from the 1984 Conference led to the production of a practical workshop report with specific recommendations for the integration of risk reduction in the development process.

These events, along with the results of the 1991 survey, influenced the development of the CDMP, which was developed to serve as an initial vehicle to address the critical issues identified in the survey. The concept was that the CDMP would develop the necessary tools required for the integration of mitigation measures in the development process. These were seen as being hazard and vulnerability maps, building codes and guidelines for the construction of informal structures, measures to improve insurance premium pricing and the implementation of measures to improve the skills of artisans involved in the construction industry. Outputs from the project would be made available to the national governments, technical institutions and community groups along with the requisite training in their use.

The importance of the community in disaster management was also recognized from the 1984 conference. Accordingly, the project incorporated a series of community-based pilot projects. The replication of these activities, which included the retrofitting of informal structures and training of local artisans in safe construction, would later be undertaken in two countries.

The development of project output activities would be done in a collaborative manner with local specialists and external professionals to ensure that the necessary knowledge transfer was achieved.

In addition to the tools, the project also encouraged the development of working alliances with the insurance sector as this was considered important in ensuring the future support for project outputs. It was felt that the project outputs of hazard assessment and risk mapping, construction technology transfer and risk audits would feed into a database that could be used by the sector linking risk information to the pricing of premiums. By advocating this approach, it was felt that insurance cost could serve as an incentive/disincentive tool to ensure compliance with local building codes and development standards. The development of this component was therefore undertaken jointly with the Insurance College of Jamaica, the Insurance Association of the Caribbean and specialist personnel brought in by the project.

3. Implementation

The CDMP was presented by the RHUDO/CAR to the OFDA in March 1992. At that time, it was agreed that a sum of US$5 million would be set aside from the international disaster assistance account for its implementation. The anticipated method for implementing the CDMP through the regional USAID offices suffered an early setback with the closure of several USAID offices in the early 1990s. This necessitated the development of a different approach in implementing the project.

In keeping with this directive, a formal contract was entered into with the Organization of American States to serve as the implementing agency. Implementation of the project commenced in September 1993 with the preparation of storm surge maps for selected coastal towns in Jamaica. The project officially ended in 1999.

The success of the CDMP is manifested in the impact it has had on the work activities and plans of agencies such as the Caribbean Development Bank, the Caribbean Disaster and Emergency Management Agency and the various disaster management agencies. The project spearheaded the development of hazard maps, disaster mitigation policies and the integration of disaster risk reduction measures in the development process. Its success is further attested to by the widespread impact of its activities that expanded from the seven original countries to 16 that became involved in the demonstration project or participated in information exchange (see www.oas.org/cdmp/document/final/oasfinal.htm).

4. Lessons learnt

In retrospect, while both the Ocho Rios Conference and the Port-of-Spain Workshop inspired the development of the CDMP, this essentially resulted in an overemphasis on the implementation of technical solutions. As noted in the Organization of American States Final Report on the CDMP, 'the tools/approaches selected to address the chosen issues also varied according to their applicability to the situation at hand'. Generally, they consisted of:

- technical assistance
- technology transfer
- information exchange/dissemination
- demonstrations
- training
- problem-focused studies
- monitoring and evaluation
- counterpart investments (see www.oas.org/cdmp/document/final/oasfinal.htm).

This focus on technical solutions may have inadvertently militated against the overall success of the project as insufficient attention was paid to the development of an overarching mitigation policy. Such a policy would have resulted in countries developing at an early stage a cohesive approach to risk reduction. The effect of this lack of integration was further noted in the 'Hurricane Ivan Recovery Program Intermediate Results Reports and Evaluation, PADCO/AECOM, 2006', which noted that 'the systematic planning and mitigation of disasters has still not been fully integrated into the development planning process. These institutional arrangements must be in place for the effective implementation of disaster risk reduction measures in physical development projects' (PADCO/AECOM 2006, p. 29).

A failure to fully appreciate the social context such as the limited effectiveness or control of the planning process by planners also affected the effectiveness of the project. This was also noted in the PADCO report, which stated that, 'Safe construction initiatives have succeeded in improving building techniques, but many informal builders are still putting up unreinforced structures with low resistance to natural shocks. On most islands, formal sector planning and development control systems are not sufficiently articulated to prevent the spread of squatter settlements to environmentally risky areas such as steep slopes and floodplains' (PADCO/AECOM 2006, p. 21).

The development of overarching policies and a better understanding of the social context undoubtedly would have increased the effectiveness of the project. Other factors such as the limited absorption capacity at the national level also played an important role in determining the pace and level at which activities could be implemented. This realization would later guide the development of disaster risk reduction activities in the Eastern Caribbean under the USAID-sponsored Caribbean Open Trade Support (COTS) project, which, though regional in scope covering the Organization of Eastern Caribbean States islands, focused on the development of island-specific projects commensurate with local absorption levels.

The disaster risk reduction component of the COTS project addressed the issue of integrating risk reduction within the development process. At the regional level, the project developed a Vulnerability Assessment Benchmarking Tool; training in business continuity plans for public and private sector entities was also conducted at the regional level. At the individual country level, coastal vulnerability assessments were conducted for Dominica and St Vincent; additionally, a multi-hazard map was produced for Dominica along with two integrated development plans. The Terms of Reference for both plans were produced from a disaster management perspective and successfully integrated risk reduction activities as lead elements. Other activities undertaken included training in safe construction, and site inspection procedures were conducted in Antigua and Barbuda along with the development of a framework for effectively integrating risk reduction in the development planning process. In Barbuda, artificial sand dunes were constructed to protect the village of Codrington from storm surges.

This breaking down of a regional project into discrete country-specific projects greatly enhanced the success of the COTS project.

5. Conclusion

The CDMP was developed based on the common perception of needs within the region as articulated at the Ocho Rios Conference and the Trinidad Workshop. These needs were interpreted as being physical in nature and therefore required physical solutions. Consequently, not enough attention was paid to the social environment in which activities were to be carried out. The mandate of the project, which also stressed the development of 'practical tools' and not studies, served to guide the project into a technical/practical approach. This perhaps limited the effectiveness of the project in terms of sustainability at the national level.

Sustainability was, however, achieved at the regional level as the CDMP influenced the work programme of CDEMA, the Caribbean Development Bank and the World Bank. For CDEMA, it laid the foundation for the development of the DERMS project, which was instrumental in bringing together national disaster coordinators and planners to explore ways of integrating hazard and vulnerability reduction in the development planning process. The development of disaster management guidelines by the Caribbean Development Bank can also be attributed to the CDMP; so too can the development of mitigation policies and plans by several national disaster offices.

Despite its shortcomings in terms of the social/psychological context, the CDMP represents a significant development in disaster management within the region.

References

Jovel, R., 1982. *Natural Disasters and Their Impacts on the Social and Economic Development of Central America and the Caribbean*. International Congress on Urban Emergencies, Cancun, Mexico.

International Federation of Red Cross and Red Crescent Societies, 1997. *World Disaster Report*, 81.

PADCO/AECOM, 2006. *Hurricane Ivan Recovery Program Intermediate Results Reports and Evaluation*.

Pan Caribbean Disaster Preparedness and Prevention Project, 1985. *Report of the Pan-Caribbean Workshop on Design Mechanisms for Reducing Vulnerability to Natural Disasters*. Kapok Hotel, Port-of-Spain Trinidad.

ENVIRONMENTAL HAZARDS
Human and Policy Dimensions

Then and now: A 30-year journey from the 'leading edge'

ELEANOR B. JONES*

Environmental Solutions Ltd, 20 West Kings House Road, Kingston 10, Jamaica

A fledgling disaster management organization in 1984, Jamaica's Office of Disaster Preparedness (ODP) hosted the pioneering international disaster mitigation conference to share information and to contemplate how vulnerability to disasters from natural hazards could be reduced. Over the period 1979–2009, the overall focus of the agency was response, given the plethora of extreme events, and the priority of the Government of Jamaica. The organization was associated with several efforts at disaster mitigation, but the need for building the agency's capacity to lead national efforts to integrate mitigation planning into development did not seem to gain traction among the policy makers. Growing losses since 2004, the evidence of climate change and the need for adaptation, and the regional and international disaster risk management agenda have brought attention to the need for review and development of Jamaica's disaster risk management capacity. This paper examines highlights of the 30-year journey of disaster management in Jamaica, and highlights proposals for strengthening the national framework and the organizational structure of ODPEM.

Keywords: comprehensive disaster management; flood; hurricane; Jamaica; training

> ...This weakness continues in spite of the history of disaster events and the increased vulnerability due to population growth, intensive urbanisation, development of marginal lands and lack of coherent warning systems. Building an effective disaster preparedness capability is hence seen as an integral part of the regional development strategy... (PCDPPP, 1983).

Building a disaster-resilient nation through a 'world class agency' was the vision of Jamaica's Office of Disaster Preparedness and Management (ODPEM) is indicated in 2009. The agency articulated its commitment to leading the process of reducing the impact of disasters on Jamaica through comprehensive disaster management (CDM), as outlined by CDERA (CDERA, 2002, 2007).

In this the 30th year of the disaster management institution, Jamaica's disaster management framework is undergoing a review towards identifying opportunities for improvement in the context of CDM. This is significant in that the journey of ODPEM over the past three decades has moved from the 'leading edge' in the Caribbean, through 'holding on', to this stage where the country is now being led in an effort to deepen and strengthen its capacity to manage its vulnerabilities, and to reduce the saga of losses and disruption of livelihoods, social systems and economic development. Buffeted by extreme hydro-meteorological events in particular, and cognisant of the heightened threat of a low-frequency/'unknown' probability/high-consequence earthquake event (Mann et al., 2008) the government has sought international assistance from the InterAmerican Development Bank in the process of change.

A central consideration for the framework review and organizational strengthening are the

■ **Email:* b.eleanor@gmail.com

doi:10.3763/ehaz.2011.0004 ISSN: 1747-7891 (print), 1878-0059 (online) www.earthscan.co.uk/journals/ehaz

reported findings of a 2008/2009 National Disaster Preparedness Audit. The audit report (ODPEM, 2009), while acknowledging some strides in emergency preparedness and operational capability, identified:

> ... salient weaknesses in the areas of administration, hazard risk reduction, and communication. The Assessment revealed that the following areas required further attention:
>
> - *Resource Management* – the availability of critical human and physical resources required in Disaster Response.
> - *Hazard Identification* and *Risk Assessment.*
> - *Logistics and Facilities* – the essential facilities and services required to support emergency management activities.
>
> The assessment further revealed that the national machinery had not kept pace with the seeming increase in severity and frequency of natural disasters. The special emphasis being placed on mitigation by the ODPEM and the expressed overall strategic priority of the Office of the Prime Minister has not received the same level of priority in the operational plans and policies of line ministries and agencies (Jackson, 2009).

These are interesting findings given the observations quoted in the statement from the Pan Caribbean Disaster Preparedness and Prevention Project (PCDPPP) in 1983 as presented above. What has happened along the way? What has been the impact of events, international conferences and agreements, exchanges and project interventions? Have issues of governance and resource allocation stymied the reach of the 'professional drive' of the early 1980s?

1. Professional awakening – 'leading edge' of the 1980s

'Old thinking' on disasters and hazards began to be washed up in 1979, following extreme flooding in western Jamaica. The population was roused by extreme rainfall from a tropical system that led to the creation of flooded karstic basins, massive debris flows, creation of new and deepening of old drainage ways, formation of new landforms, devastation of agriculture, destruction of roads and massive social dislocation. The largest of the karstic lakes in the Newmarket basin remained for 9 months.

Earth scientists in Jamaica saw, for perhaps the first time in their professional lives, a 'catastrophic geomorphic event'. A sea change was about to take place in the earth science community regarding interest in the behaviour of natural hazards and their interface with human settlements. Several disaster risk management professionals practising in the Caribbean today were spawned directly and indirectly from the interest stimulated by that event, others following closely behind.

A diagnostic review of the Jamaican event (as well as events in St Vincent and Dominica also in 1979) revealed 'national weaknesses in communication, and coordinating systems, and a general lack of national policies for integrating disaster prevention and preparedness mechanisms in national economic planning' (PCDPPP, 1983). International expertise for the review in Jamaica was facilitated by a combination of international agencies and NGOs – United Nations Disaster Relief Organization (UNDRO), United States Agency for International Development Office of Foreign Disaster Assistance (USAID-OFDA), International Red Cross and the Jamaica Council of Churches. The review was coordinated on the ground in Jamaica by the Jamaica Red Cross (F. McDonald, personal communication, 30 July 2010). The findings of the review formed the basis for the establishment of the Office of Disaster Preparedness (ODP) in July 1980, as the Ministry of the Public Service adopted the findings within the systems of the Government of Jamaica. The ODP was first placed within the Office of the Prime Minister, which signalled the importance placed on the matter of disaster preparedness at the time.

Hurricane Allen, one of the largest Atlantic hurricanes in history, passed to the north of

Jamaica on 6 August 1980 generating a storm surge that wreaked havoc on agriculture, coastal settlements and coral reefs along the east and north coasts of Jamaica. Yet again, scientists were able to observe the effect of hazards on environmental systems, and the interaction of hazards and vulnerable settlements. The ODP began to build a focus on the capacity to deliver on its mandate of response, and research and planning were integrated into that portfolio.

New approaches began to emerge, and took hold among the scientific community and practitioners, but did not truly sustain the imagination of the political directorate. The ODP was shifted to the short-lived Ministry of Environment, Science and Technology, and thereafter was passed through several different Ministries of Government and without the resources required to build its capacity to handle catastrophic events. In spite of the occurrence of other major hazards during the 1980s – flooding from outer bands of Hurricane Kate in 1985, the devastating floods of June 1986, and the catastrophic Hurricane Gilbert of 1988 – interest by the government structure in strengthening response beyond immediate relief was not sustained.

Jamaica participated in the PCDPPP, which was formulated by the Caribbean Community (CARICOM) Secretariat to help build the capacity of the Caribbean to respond to natural hazard events, and to assess vulnerabilities towards prevention of disasters. CARICOM, UNDRO, League of Red Cross Societies and Pan American Health Organization/World Health Organization (PAHO/WHO) were the executing agencies of the programme, providing multidisciplinary expertise to help fulfil the organization's mandate.

In the English-speaking Caribbean, Jamaica was the leader in institutional arrangements for disaster management, having been the first country to establish a national disaster office. However, the real leadership came from the international community and voluntary sector, which had an interest in reducing loss, suffering and dislocation, and from the earth scientists who were beginning to seek answers and to share information towards building a knowledge base.

1.1. Defining events

Disaster management teaching was introduced into geomorphology courses at the University of the West Indies, and students were encouraged to undertake research to develop parish hazard profiles that were to assist building the database for ODP's work. USAID supported damage assessment and development of a disaster mitigation strategy from 1983–1984, and the signature International Conference on Hazard Mitigation was convened in Ocho Rios in 1984. 'Mitigation' had reared its head in Jamaica, and vulnerability as the salient contributor to disaster was beginning to receive some attention.

The record of the 1980s was busy with defining events. The June 1986 floods mentioned above revealed the inextricable links between extreme rainfall and environmental degradation, poor planning, inadequate design and maintenance of infrastructure, inter alia, and the deleterious effects were documented in a video entitled 'The Case for Mitigation', the storyboard for which remains highly relevant in 2010.

Some international exchange programmes facilitated capacity building among first responders and within the private sector. A flood plain mapping project for two vulnerable watersheds was implemented through the Water Resources Authority with the assistance of the World Meteorological Organization. The intent was to identify flood zones to guide planning and development. A study on the vulnerability reduction assets of vernacular housing was undertaken by Fred Cuny of Intertect, USA, and documentation was shared in the building industry. A training programme in hazard-resistant construction techniques for builders was designed and implemented with resource material for general distribution. The College of Arts, Science and Technology (CAST), which is now the University of Technology, introduced the teaching of disaster management in the physical planning technology course and graduates were recruited by the ODP.

Planners and engineers, architects and risk management professionals assembled in Trinidad to discuss the integration of natural hazard vulnerability reduction considerations into the respective professional endeavours, and discussion on development of a Caribbean Uniform Building Code (CuBIC) was initiated.

Hurricane Gilbert in 1988, a Category 4 storm when it travelled across Jamaica, laid bare the significant vulnerabilities of the island. Infrastructure failure, roof lifting and building collapse were among the many catastrophes experienced, particularly in the Kingston Metropolitan area and other urban centres. Assessment of causative factors revealed the inadequacy of prevention and mitigation measures in the building and construction industry. Destruction of agriculture was widespread, and the effort towards reconstruction and recovery was stymied by a national disaster framework that proved inadequate to handle the scale of the disaster.

The organization had been renamed Office of Disaster Preparedness and Relief Coordination to account for the amalgamation of the island's Relief Dept that had been established after Hurricane Charlie in 1951. Relief coordination was emphasized along with disaster response.

2. What then of the 1990s?

The disaster management framework went into a 'holding mode', with response-orientation continuing as its major focus. The subsequent name change to Office of Disaster Preparedness and Emergency Management (ODPEM) in 1993, following promulgation of the Disaster Preparedness Act, retained the focus on response. The national disaster plan with hazard-specific subcomponents was also developed during this decade.

Several flood events caused infrastructure damage, but the reasons for failure were not fully assessed or integrated into reconstruction. Parish disaster planning was implemented in an effort to build capacity at the community level, but local government resource constraints stymied the outcomes.

Governance for disaster management was generally weak in that the lead agency had been shifted between several ministries from the 1980s. There was a succession of directors for many years, and some had no background in disaster management. There appeared to be no real integration into the development agenda. However, the ODP progressively expanded space to accommodate its operations, and staffing remained at the highest level in the region, although inadequate to meet the disaster risk management requirements.

Efforts to introduce a comprehensive approach to disaster management were introduced in the late 1990s to early 2000s, and the endorsement of the CARICOM Caribbean CDM Strategy has become the operational framework for ODPEM (CDERA, 2002, 2007).

3. How different is the 21st century?

Within the past decade, 2000–2010, Jamaica suffered several extreme flood events that wreaked havoc on infrastructure. In 2004 the island was struck by Hurricanes Charley in June and Ivan in September. Charley caused severe damage to agriculture in the southwestern parishes, and Ivan caused wind and flood damage in most parishes across the island. Severe erosion within the coastal zone and in the watersheds caused losses to housing and infrastructure. The high exposure to risk and the lack of awareness of vulnerable populations was marked. A national review of the response to Ivan was undertaken through stakeholder consultations and reporting from key sectors and committees. The findings highlighted shortcomings at the parish level and in the national response system, including utilities, health and shelters.

In 2005 Hurricanes Dennis, Emily and Wilma passed within sufficiently close proximity to the island to discharge tons of rainfall that in turn caused further damage to infrastructure and livelihoods. Hurricane Dean damaged the southern parishes in 2007. The only road to the Norman Manley airport in Kingston was overtopped by storm surge as the dunes and coastal defence works had not been restored/enhanced following the destruction by Hurricane Ivan.

In 2008 TS Gustav again unloaded heavy rainfall in many parts of the island, causing severe damage and dislocation to livelihoods and infrastructure. The exposure of vulnerable populations and the limitations of the national response mechanism were repeatedly highlighted.

The experience with Hurricane Dean highlighted needs in the areas of shelter availability and management, damage assessment and recovery. Weaknesses in the welfare assessment procedures again emerged (as after Ivan), (Jackson, 2008) and the need for a clearly articulated recovery plan was highlighted (Jackson and Clarke, 2008) .

The disaster record of the period 1979–2008 is instructive. Seven of the top 10 disasters in Jamaica's recorded disaster history in terms of persons affected occurred during this period (Table 1).

Nine of the top 10 disasters in terms of economic loss also occurred during this period, as illustrated in Table 2.

4. 2008 – A New Thrust?

In the light of the disaster experience of 2007, 2008 and the previous years, and under new leadership, ODPEM reviewed and revised its mission and goals. In 2008 the overarching goals were articulated as:

- sustainable development in Jamaica enhanced through CDM and
- ODPEM's institutional capacity strengthened and corporate profile raised.

The Director General of ODPEM in his report to the National Disaster Committee (NDC) in 2009 identified some key issues impinging on the effectiveness of the DRM agenda in Jamaica:

- climate change adaptation
- land tenure and the implications for vulnerability reduction
- institutionalization of recovery.

In 2009, the Government of Jamaica (GOJ) entered into a technical cooperation agreement with the IDB to strengthen the National Disaster Risk Management capacity (Table 3).

TABLE 1 Top 10 natural disasters in Jamaica for the period 1900–2010 sorted by numbers of total affected people

Disaster	Date	No. total affected
Storm	12-Sep-1988	810,000
Flood	21-May-1991	551,340
Storm	11-Sep-2004	350,000
Flood	12-Jun-1979	210,000
Drought	Jan-1968	100,000
Storm	18-Nov-1912	94,820
Earthquake (seismic activity)	14-Jan-1907	90,000
Flood	25-Apr-1979	40,000
Flood	15-May-1986	40,000
Storm	20-Aug-2007	33,188

Source: CRED (2009).

TABLE 2 Top 10 natural disasters in Jamaica for the period 1900–2010 sorted by economic damage costs

Disaster	Date	Damage (000 US$)
Storm	12-Sep-1988	1,000,000
Storm	11-Sep-2004	595,000
Storm	13-Aug-2004	300,000
Storm	20-Aug-2007	300,000
Flood	15-May-1986	76,000
Storm	28-Aug-2008	66,198
Storm	31-Jul-1980	64,000
Storm	17-Aug-1951	56,000
Storm	6-Nov-2001	55,487
Flood	Nov-1987	31,000

Source: CRED (2009).

5. Synergy with the national development strategy – Vision 2030

The disaster risk management agenda should be synchronized with the long-term National

TABLE 3 Highlights of the 30-year journey

1979	1980–1989	1990–1999	2000–2010
Newmarket floods – 1979	ODP established July 1980	ODP main focus response	Several damaging flood events, hurricanes
Review of response to and impact of 1979 events in St Vincent, Dominica and Jamaica	Response, research and planning, and library were the three initial departments	Organization in 'holding mode'. Succession of directors – many no background in DM	2001–2002 – extreme flooding – infrastructure damage
Government of Jamaica responds to findings and recommendations to set the stage for the establishment of ODP	Hurricane Allen August 1980 Natural hazard vulnerability study introduced to geomorphology teaching and student research at UWI – 1980–1981 Natural hazard mitigation strategy for Jamaica developed 1983–1984 with support from USAID ODP and GOJ host International Natural Hazard Mitigation Conference – Ocho Rios, 1984 Floodplain mapping project – two watersheds Vernacular Housing Study – Traditional designs and vulnerability reduction Training of builders Feeder bands from Hurricane Kate – 1985 – Flooding Southern parishes June 1986 – Extreme flood damage – Southern parishes Hurricane Gilbert September 1988 – severe damage, destruction and loss islandwide Disaster management introduced to physical planning technology course at CAST – graduates recruited to staff ODP Engineers and planners from Jamaica and other territories met in Trinidad to discuss mitigation and development of CuBIC	Flooding – 1991–1992 – Port Antonio area 1993 – Earthquake Kingston Disaster Management Act – 1993. Name change to ODPEM Caribbean Disaster Mitigation Project produced some micro-zonation maps for Jamaica – storm surge, landslides, Kingston Metropolitan area – urban hazards Parish Disaster Committees established	2004 – Hurricanes Charley, Ivan 2005 – Hurricanes Dennis, Emily, Wilma produced flooding and landslides – damaged infrastructure 2007 – Hurricane Dean – critical infrastructure damage 2008 – Tropical Storm Gustav – damage to watershed, infrastructure and agricultural livelihoods Review and adoption of CDM approach Jamaica hosts regional CDM conference Internationally funded project support for work of ODPEM Strategic plan – visioning Review of disaster management framework, and legal and organizational structure towards strengthening disaster risk management in the country

Development Plan of the Government of Jamaica, Vision 2030, which is intended to redefine the strategic direction for Jamaica towards sustainable prosperity. The vision is for Jamaica to move towards development of cultural, human, knowledge and institutional capital stocks so as to attain higher stages of development by 2030. One of the seven guiding principles on which the plan is based is sustainability (economic, social and environmental).

Four national goals have been identified, and the fourth of these is *Jamaica has a healthy natural environment*. Three national outcomes are linked to this goal:

1. #13. sustainable management and use of environmental and natural resources;
2. #14. hazard risk reduction and adaptation to climate change; and
3. #15. sustainable urban and rural development.

It is fair to say that several of the other goals and outcomes are either directly or indirectly linked to reduction of loss from the occurrence of extreme natural or technological events.

The national strategies related to reducing hazard risk and adapting to climate change are as follows:

14-1. Improve resilience to all forms of hazards.
14-2. Improve emergency response capability.
14-3. Develop measures to adapt to climate change.
14-4. Contribute to the effort to reduce the global rate of climate change.

Jamaica's development patterns and population characteristics are increasing its vulnerability to both existing and emerging threats. With the exception of Haiti, Jamaica is the largest of the CARICOM partners within the CDEMA framework of participating states. Urbanization and several associated parameters of the urban process constitute a major aspect of the island's vulnerability to both natural and technological hazards. Seismic risk, urban search and rescue, mass casualty management, evacuation and shelter are some of the salient concerns in that regard. Some coastal towns, which are included in the IDB Technical Cooperation Programme, are being assessed so as to guide development of the respective disaster management plans. The methodology and approach will be expanded to other areas, some of which are already being assessed under other programmes.

The importance of an effectively designed and implemented disaster risk management programme is germane to sustainable prosperity in Jamaica. The proposed enhancement programme for ODPEM must therefore take cognisance of the objectives of Vision 2030 as articulated, and the recommendations must be so informed.

6. The legal underpinning

The Disaster Management Act of 1993 was revised to be superseded by the Disaster Management and Mitigation Act of 2008. A legislative review was conducted on the 2008 Act and several amendments were made. Some new sections have been added, Shelter By-laws prepared and the title changed to Disaster Risk Management Act 2009 (McCalla, 2009).

The new law is geared towards strengthening areas in need of immediate attention: legal recognition of existing organizational structures such as the NDC, the Parish Disaster Committees and the Zonal Committees, with their roles and functions clearly established; the legal establishment of a National Disaster Fund; legal evacuation of persons identified to be at risk as a prevention measure; and the identification and description of high-risk areas as 'specifically vulnerable areas' and the necessary action to be taken in respect of such areas.

The 2009 Draft Act provides the key legal institutional framework to prescribe:

- acceptance of the current Revised Law with some modification
- the Disaster Council (now recognized by law)
- the Board for ODPEM as a statutory body

- an Executive Director and staff
- Parish Disaster Committee Structure.

The law provides for flexibility in terms of operational structure and staffing.

7. Revisiting the existing national disaster framework

Provisions of the existing national disaster management framework are seriously out of date with respect to the several regional and international agreements and protocols that have been agreed upon, and to which Jamaica is signatory. The UN Framework Convention on Climate Change, the Hyogo Framework for Action, the Millennium Development Goals and the Caribbean CDM Strategy are some examples (UNWCDR, 2009).

Further, as shown above, the island has been bombarded by several extreme events that have wreaked havoc on people, land, livelihood and infrastructure. Macroeconomic implications have not been fully assimilated, and methodologies for complete assessments of financial and economic losses at the micro levels are not generally available to be applied. Key lessons learnt from Jamaica, its neighbours in the Caribbean and the experience of other international Small Island Developing States should serve to inform the required modifications to the system of managing disaster risk.

'Authority to coordinate' looms large when there is a catastrophic event, and the recent experience with the earthquake event in Haiti in January 2010 is an astounding reminder of the many contingencies that require planning for humanitarian aid and continuity of government and civil society.

The framework as currently obtained includes national-, parish- and community-level functions and responsibilities, as well as regional responsibility under the CARICOM Response Agreement. Jamaica serves as a focal point for the Western Caribbean Region.

The national-level responsibility is primarily a central government function headed by the NDC, which is chaired by the Prime Minister of Jamaica and comprises a wide range of stakeholders from public (national and local), private, non-governmental and civil sectors of society. The remit of the NDC is to provide policy direction and oversight for DM in Jamaica. To date, there has been no legal basis for the NDC.

The committee was originally conceptualized following a series of dislocating climatic events in the late 1970s; it reportedly lay relatively dormant during the 1980s, only to be resuscitated following the devastation of the island caused by Hurricane Gilbert in 1988. The NDC currently meets once a year at the start of the hurricane season, and receives reports on preparedness and policy requirements. This practice is deemed inadequate as it demonstrates the dominant focus on preparedness for hurricanes in the multi-hazard environment of Jamaica. Further, the format of the meeting as currently constituted does not lend to effective dialogue and strategic discussion. The opportunity for strategic exchange at the level of national leadership for disaster risk management cannot be met by this single meeting.

Regional and parish committees represent the second tier of the framework. Parish Councils (PCs), the accountable local government agencies at the parish level, are ultimately responsible for disaster management in the parish and the mandate is discharged through the Parish Disaster Committee. Four regional coordinators are employed by ODPEM to monitor disaster management at the parish level and to assist the parishes in developing and executing the parish disaster plan. A Parish Emergency Operations Centre is also established at the PC. The Parish Disaster Coordinators (PDCs) report to the PC.

The community, the third tier of the framework, is linked through zonal groupings to ODPEM. Building community resilience is a stated objective of ODPEM's programme, given Jamaica's disaster experience and projected needs, and in keeping with the commitment to the Hyogo Framework. A 'community resilience-building project' is being undertaken primarily through the support of Canadian International Development Agency (CIDA), and there are other community-based initiatives supporting the strategic direction.

The subcommittees of the NDC were intended to link the operations of ODPEM with the policy-making body, but they reportedly do not function well. In any event, the subcommittees as currently constituted represent selected thematic areas primarily with a response focus as opposed to the holistic needs of a comprehensive disaster risk management approach.

In summary, the framework that has been employed over the years for the management of disaster risk in Jamaica is inadequate to handle the existing as well as emerging threats with the holistic approach that is required. A governance structure is needed to facilitate mainstreaming/integration of risk management initiatives, and the structure of ODPEM needs to be strengthened with the technical expertise and organization that address the disaster risk management continuum in terms of prevention and mitigation, preparedness and response, recovery and rehabilitation, but also the supporting focus on risk identification, risk transfer and risk reduction. The name of ODPEM needs to be changed to the National Disaster Risk Management Authority (NDRMA), and the legislation should reflect this.

8. Proposed national framework

Under the proposed change in the framework, the National Disaster Risk Management Council (NDRMC), which is provided for in the revised legislation awaiting enactment, would be chair by the Prime Minister and would meet once per year to receive the annual report on the status of national disaster risk management programming, which would in turn be tested in Parliament. The next tier would be the National Disaster Risk Management Committee (NDRMCo) to be chaired by the parent minister, with permanent secretaries of core ministries as members. The NDRMCo would meet and would be charged with the following responsibilities:

- To review in collaboration with NDRMA the state of DRM of the country and recommend areas to be addressed.
- To monitor the progress of the respective ministries in the development of ministry and sector disaster plans including government continuity.
- To agree on annual disaster risk management budget allocation for ministries – part of the corporate planning process.
- To assist the Prime Minister to brief parliament on the state of preparedness of the country and proposed actions to address gaps.
- To collaborate with NDRMA on the development of annual work programmes, budget and linkages with the respective sector ministries and Emergency Support Functions (ESFs).

This committee would operate at the policy-making level and would work in collaboration with the NDRMA Board for supporting delivery of the mandate of the leadership of the country's disaster management imperative. The idea is to have an NDRM Committee that is more interactive with the Director General and the NDRMA Team, so as to be able to assist NDRMA and its partner agencies to garner and/or leverage the resources necessary to further disaster risk management in the country. Further, the permanent secretaries as head of their respective ministries will help to drive the mainstreaming process as their ministry mandates would need to include DRM with the requisite budget lines, and they would have oversight responsibility for monitoring progress. It is proposed that these provisions be included in the revised legislation.

It is recommended that the revised structure link the NDRMC to the Strategic Planning function of the Office of the Prime Minister and Cabinet Office so as to facilitate mainstreaming of disaster risk management imperatives into strategic planning for the country.

The parish-level responsibilities within the PCs will continue to be led by the respective Parish Disaster Committee. Within the PCs, Directors of Planning should be trained in the requirements and application of DRM so as to ensure integration of CDM into the parish planning and enforcement agenda. Decentralization of mitigation and response functions are particularly important for loss reduction at the parish level.

The system of regional coordinators employed by NDRMA will be expanded to include three additional officers to provide for more effective allocations to parish groupings on the basis of population size and vulnerabilities. The major urban centres of the Kingston Metropolitan area, and those within St Catherine and St James, will receive particular attention.

Zonal groupings of communities will remain, but will achieve greater attention as the community disaster resilience building programme is expanded.

The proposed functional relationships among the key leaders of the disaster risk management framework are illustrated in Figure 1.

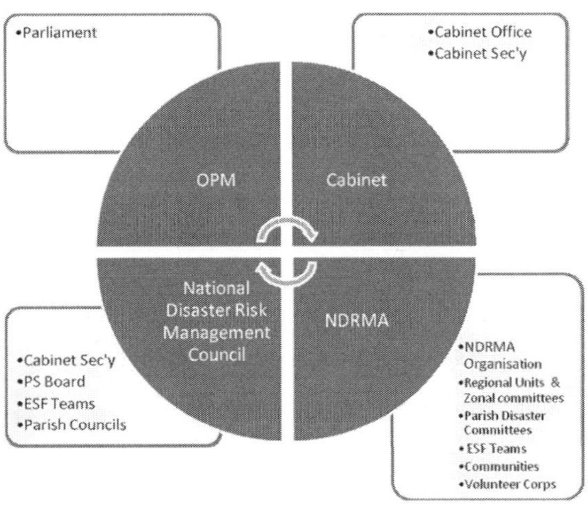

FIGURE 1 National disaster risk management framework (proposed)

9. The National Disaster Risk Management Agency (NDRMA/ODPEM)

As indicated above, it is recommended that ODPEM should be renamed NDRMA to reflect the revised focus on disaster risk management. The agency has outlined a Strategic Plan for the period 2008–2011 and the priority outcomes identified are:

- sustainable development in Jamaica enhanced through comprehensive disaster management;
- strengthening ODPEM's institutional capacity and raising the corporate profile;
- creating a motivating, challenging and safe work environment for ODPEM employees.

ODPEM has evolved since the establishment of the agency as ODP in 1980. The national, global and regional DRM agendas have changed significantly since the 1980s as a function of events, lessons learnt and ongoing research. However, it is significant that ODPEM's human resources structure has not been reviewed since the 1980s and there has been no reclassification. The salary structure and organization are therefore woefully out of date. More importantly, there are no posts for the programmes that need to be undertaken, and several 'acting' positions do an injustice to the persons so appointed as well as to the jobs that need to be done. The staff complement and structure militates against a fully functioning portfolio at the national, parish or community level.

In the new thrust towards building national and community resilience, the pursuit of sustainable development through enhanced comprehensive disaster management, and sustainable prosperity as articulated in Vision 2030, ODPEM is required to strengthen its internal capability while leading a significant shift in the wider national environment. This includes the state, its agencies, public and private institutions, and civil society organizations. In order to deliver on this imperative, ODPEM's organizational structure and modus operandi must change substantially within the short to medium term. The required knowledge base, skill sets and technological resources must be accessed and/or developed as the organization seeks to build the partnerships and alliances at the regional, international, national and local levels for delivery of a results-based comprehensive disaster risk management programming framework, and the CDEMA Regional Response Mechanism. Strategic planning is imperative, and a monitoring framework is essential to guide the performance evaluation of results, as well as project/programme funding that has been of critical importance to ODPEM given the budgetary constraints of the national treasury.

It is proposed that the ODPEM organization therefore focus on the strategic as well as operational functions. The Director General will have primary responsibility for the strategic functions at the executive level of the organization. In that regard, he will have direct oversight of strategic planning, monitoring and corporate communications, three significant strategic planks for going forward. In addition, the Internal Auditor will report directly to him/her.

The Corporate Communications arm should be the communications interface between ODPEM and its several publics. The unit will be responsible for the development and maintenance of a structured communications protocol, crisis communications plans and public relations messaging.

The DG should be supported by a deputy who will have responsibility for the two main divisions of the organization:

1. Disaster Risk Management and Technical Services (DRMTS) and
2. Administration and Corporate Affairs (ACA).

The DRMTS should be underpinned by strong research, documentation and information technology functions that will guide preparedness and response, mitigation and planning, and knowledge management. The ACA Division should be led by a Corporate Planner. A strong financial management portfolio under the leadership of a Chief Financial Officer who will have responsibility for the organization's budgeting for results and property management functions should be included in this division.

ODPEM has several needs. One of these is a research team with the requisite skills in social, economic and physical sciences, and planning disciplines. Vulnerability assessments by parish and by community need to be conducted and to be followed by development of risk profiles and vulnerability ranking for vulnerable areas and communities. These will pull together poverty data, history of events, critical facilities, economic profiles and other relevant data to inform the disaster risk management agenda.

The preparedness and emergency response functions need to be strengthened to facilitate building disaster resilience at the national, parish and community levels. This aspect of the agency's mandate is critical not only for humanitarian assistance, but importantly for damage assessment to inform mitigation needs for restoration and recovery. Of additional consideration is ODPEM's role as a regional focal point for the Caribbean western region under the CDEMA agreement. The Preparedness and Emergency Response Operations Division should build the capacity to provide the requisite leadership.

The national disaster plan structure requires an Emergency Operating Division that does not now exist. Administrative and programme persons are assigned if needed in case of an emergency. Emergency service representatives are needed to function independently and actively in-house.

10. Prospect

The Government of Jamaica has embarked on a 'transformation exercise' within the public service, with the objective of improving efficiencies while reducing costs. The transformation is expected to take account of the government's strategic vision, and based on Vision 2030, disaster risk reduction is integral to building the country's sustainable prosperity. The proposed revision, which is yet to have the benefit of national consultations and parliamentary debate, places the disaster management function with other agencies responsible for sustainable development. In that regard, the opportunity for interaction and collaboration towards integration of disaster risk management could be enhanced.

The proposed reorganization of ODPEM, including the renaming of the organization to

NDRMA, and deepening of technical capacity to reflect the DRM mandate, will need to be considered by the entity for public service strategic direction and administration, and then recommended to the Cabinet and Parliament of the Government for adoption. The legislation will need to be amended to reflect the proposed changes and to establish the legal basis for recommended action.

The 'leading edge' of the pioneering initiatives of the 1980s should be borne out by a national drive to replace expedience with due diligence and informed approaches to integrating disaster risk management into the development agenda. The 'weakness' cannot persist as it was reported in 1983, and in 2010. Damage and loss reduction and a disaster-resilient nation must be the guiding mantra for at least the next three decades. A review in 2040 should reflect real achievement.

References

CDERA, 2002. *Comprehensive Disaster Management. Strategy and Program Framework. 2002–2007*. DERA, Barbados.

CDERA, 2007. *Comprehensive Disaster Management. Strategy and Program Framework. 2007–2012*. DERA, Barbados.

CRED, 2009. *EM-DAT Top 10 Natural Disasters Jamaica*. Centre for Research on Epidemiology of Disasters, Jamaica.

Government of Jamaica (GoJ), 2002. *The Public Bodies Management and Accountability Act*. Jamaica.

Jackson, R., 2009. *The Director General's Report to the 2009 National Disaster Committee Meeting*. Office of Disaster Preparedness and Emergency Management (ODPEM), Jamaica.

Jackson, R., 2008. *Revised National Disaster Welfare Response Programme*. Office of Disaster Preparedness and Emergency Management (ODPEM), Jamaica.

Jackson, R. and Clarke, D., 2008. *Establishing an Institutional Framework for Disaster Recovery: Presentation to the Permanent Secretaries Board*. Jamaica.

Mann, P., Calais, E., Demets, C., Prentice, C S. and Wiggins-Grandison, M., 2008. Enriquillo-Plantain garden strike-slip fault zone: a major seismic hazard affecting Dominican Republic, Haiti and Jamaica. Presented to the *3rd Caribbean Geological Conference*, Puerto Rico.

McCalla, W., 2009. *Review of Institutional and Legal Framework Development of Subsidiary Legislation: Final Progress Report*. Jamaica.

Office of Disaster Preparedness and Emergency Management (ODPEM). *National Disaster Organisation Assessment*. Jamaica.

PCDPPP, 1983. *Pan Caribbean Disaster Preparedness and Prevention Project*. Comprehensive Project Document. Revised and approved at the Management Committee Meeting, Barbados, April.

Planning Institute of Jamaica, 2009. *Vision 2030. National Development Plan*. Jamaica.

UNWCDR (United Nations World Conference on Disaster Reduction), 2005. *Monitoring Progress on Implementation of the Hyogo Framework for Action (HFA): HFA Monitor Template*.

ENVIRONMENTAL
HAZARDS
Human and Policy Dimensions

Revisiting community-based disaster risk management

ANDREW MASKREY*

United Nations Strategy for Disaster Reduction, International Environment House II, 7–9 Chemin de Balexert, CH 1219 Chatelaine, Geneva, Switzerland

This paper examines whether the hypothesis presented in Ocho Rios in 1984 is still valid and relevant. The first part of the paper revisits the central tenets of the 1984 paper. The second part of the paper gives a broad overview of how community-based approaches have been taken up and evolved over the last quarter-century. The third part of the paper examines, with the benefit of hindsight, to what extent the approaches advocated in Ocho Rios are still valid. Finally, the paper examines the role of community-based and local-level approaches looking into the future. The paper is written as a personal reflection, without any pretence to academic rigour and also without systematically surveying the thousands of ongoing community-based or local-level initiatives in disaster risk management, referred to in this paper as community-based disaster risk management and local-level disaster risk management, respectively.

Keywords: community-based disaster risk management; local-level initiatives in disaster risk management; Ocho Rios; risk assessment

1. Introduction

Twenty-five years ago, the paper 'Community based hazard mitigation' (Maskrey, 1984) appeared as part of the proceedings of the International Conference on Disaster Mitigation Program Implementation, held in Ocho Rios on 12–16 November 1984. Before that time, many organizations and individuals in different parts of the world had been working with local households and communities to mitigate disaster impacts (e.g. following the Guatemala earthquake in 1976, the Yemen earthquake in 1982 and the Popayan earthquake in 1983). Approaches to addressing vulnerability at the local level were also starting to emerge in the literature (Davis, 1978, Wisner, 1979; Bommer, 1985; Egas 1985; Leslie, 1986; Wilches-Chaux, 1989). However, the Ocho Rios paper was, to my knowledge, one of the first to systematically discuss approaches to engaging local communities to manage and reduce disaster risk as an object of analysis in itself.

This paper examines whether the hypothesis presented in Ocho Rios in 1984 is still valid and relevant. The first part of the paper revisits the central tenets of the 1984 paper. The second part of the paper gives a broad overview of how community-based approaches have been taken up and evolved over the last quarter-century. The third part of the paper examines, with the benefit of hindsight, to what extent the approaches advocated for in Ocho Rios are still valid. Finally, the paper examines the role of community-based and local-level approaches looking into the future. The paper is written as a personal reflection, without any pretence to academic rigour and also without systematically surveying the thousands of ongoing community-based or local-level initiatives in disaster risk management, referred to in this paper as community-based disaster risk management (CBDRM) and local-level disaster risk management (LLDRM), respectively.

*Email: maskrey@un.org

doi:10.3763/ehaz.2011.0005 ISSN: 1747-7891 (print), 1878-0059 (online) www.earthscan.co.uk/journals/ehaz

2. From somewhere along the time coast: Ocho Rios, 1984

The Ocho Rios paper itself was written, not as an academic essay, but as a personal reflection on ongoing experiences in supporting low-income urban and rural communities exposed to natural hazards in Peru. In 1983, Gilberto Romero, Jose Sato and this author had left the governmental National Institute of Urban Development (INADUR), where they had developed a study of seismic vulnerability of the Lima Metropolitan area (Maskrey and Romero, 1985), to form the Centre for Disaster Prevention and Studies (PREDES). PREDES was a non-governmental organization that is still active to this day. Unlike INADUR, which developed plans and studies for government, PREDES was formed to directly advise communities and local governments on how to manage and reduce their risks. While initially focused on communities in the landslide- and flood-prone Rimac Valley, to the east of Lima, the services of PREDES were soon in demand in other areas of Peru.

The approach of PREDES took as its starting point that not only was most disaster risk concentrated in poor rural and urban communities but that the exposure and vulnerability of these communities to natural hazards were anything but accidental. It was an inherent characteristic of a development process that generated inherent and dramatic socio-territorial inequalities, one of which was the concentration of disaster risk in poor communities. For example, the urban poor in the Rimac Valley had no alternative but to occupy hazard-prone sites with the lowest land values; self-build often insecure and vulnerable housing; and develop their neighbourhoods without the material or technological resources to reduce risks. The Ocho Rios paper described the condition of these communities as a *daily disaster*, a concept that has also been described as everyday risk (ISDR, 2009) and chronic risk (Sen, 2000). It goes on to state:

> The occurrence of so-called natural disasters is nothing more than the temporary acceleration and accentuation of this daily disaster by a given natural hazard. To explain, therefore why the destruction caused by a natural hazard becomes a grave disaster for the poor majority it is necessary to analyze the social relations of production through key factors such as: the characteristics of the production system; the internal distribution of resources; the presence of a high level of unemployment and low or non-existent salary levels.

As PREDES engaged risk-prone communities, it also became very clear, very quickly, that the government of the time seemed to have little interest in reducing the risk they faced. When disasters occurred, both government agencies as well as relief mechanisms such as the Red Cross certainly provided humanitarian assistance to those affected. The government also invested in physical mitigation measures to protect strategic infrastructure such as major highways and hydroelectric plants. But no policy, plan or programme existed to reduce the risks faced by poor communities. PREDES realized early on, therefore, that communities themselves had to create the opportunities for risk reduction. Again quoting from the Ocho Rios paper:

> Genuine hazard mitigation must therefore be directed by those who suffer the effects of the disaster and is thus, almost by definition community based hazard mitigation ... where local communities have a long history of struggle and organisation to resolve basic necessities such as drinking water, food, housing etc. hazard mitigation is nothing more than an additional activity which must be undertaken in order to survive.

The experience that emerged in the Rimac valley in Peru also indicated that while communities had to be the active subject of disaster risk management, there were very clear limits as to what they could achieve on their own. Communities were already engaged in many activities to reduce risk both at the household and communal level (an activity that has more recently been referred to as *coping* or even *spontaneous adaptation*) as part of their day-to-day struggle for social and economic well-being. Communities could do much to improve their own preparedness and response capacities or to undertake localized physical hazard mitigation measures. However, they simply did not control resources and could not influence decision-making processes in a way that could unlock access to safe land, enable the management of complex watersheds or implement the large-scale public works necessary to reduce risk. Progress, therefore, would depend on communities progressively engaging and involving supra-local actors such as local and central government agencies and non-governmental organizations (NGOs) to support their activities. As the Ocho Rios paper states:

> In most contexts and for most hazards, mitigation normally includes measures at many different levels of action, which are superimposed and co-exist simultaneously.... Community based hazard mitigation does not signify the implementation of mitigation measures exclusively at the community level of action, but the progressive activation of all the levels, starting with the community itself.

This last quote synthesizes probably the essence of CBDRM. It was certainly what the Ocho Rios paper was all about. As will be argued below, it may have proved elusive in practice. The Ocho Rios paper argued that CBDRM was neither specifically, nor particularly about implementing mitigation actions at the community level. It was about empowered communities changing roles from objects to subjects in a very dynamic process of vulnerability reduction, enabling them to negotiate resources and support from local and central governments to undertake risk management measures at all scales. In other words, linkages across scales and between civil society and government were seen to be key to unlocking the political and economic resources required in order to manage risks. The challenge, therefore, was not how to encourage communities to participate in local-level risk reduction activities. As mentioned above, most communities were already doing what they could to manage their perceived risks with the limited resources at their disposal. The real challenge, rather, was to encourage governments and other supra-local actors to participate in supporting community-driven processes. The Ocho Rios paper states:

> Community based hazard mitigation can thus be seen to have two axes: an axis of development of social, cultural and political consciousness and organization and an axis of development of specific mitigation measures. The 'art' of community based mitigation is the art of achieving a balance between these two axes. In the same way that the broad structural contradictions of the development process are reflected in the very specific problems of a small settlement, such as poor siting and vulnerable construction, the reverse is also true. The resolution of specific local problems is the entry point, which permits access to the transformation of the structural contradictions... The centre and starting point of community based hazard mitigation, where both axes meet, is the community itself, from where the process evolves in a spiral form, to progressively encompass all levels of action ... In conclusion, in community based hazard mitigation, the key issue is not how to persuade the community to participate in government-based mitigation programmes, but rather how to persuade government at all levels to participate in communities' own programmes.

That was the vision set out in Ocho Rios in 1984. The paper went on to argue for a set of enabling activities: raising awareness of risks and vulnerabilities, using techniques later popularized as Community Risk Assessment Tools (ProVention, 2007); supporting and strengthening community organization; and applying appropriate disaster risk reduction measures to resolve specific problems. It also highlighted that NGOs could and should play a key role in enabling CBDRM. Above all, the paper suggested that CBDRM could have a number of key advantages over conventional top-down disaster risk management.

Firstly, the most important of these advantages, already highlighted above, is that CBDRM, through empowering communities, had the potential to restructure the relations between civil society and the state and between the local and national scales in such a way that underlying structural vulnerabilities could be reduced over time. This idea was not fundamentally different from the rationale behind the development of social movements in other areas, such as housing, water and sanitation, education and health, as well as building on the notions of popular education expressed by Freire (1970).

Secondly, the paper argued that CBDRM was more effective because it had the potential to mobilize and unlock the application of resources and know-how at the community level, which in top-down disaster risk reduction tends to be ignored or left aside. In other words, the cost of disaster risk reduction could be dramatically reduced through unlocking local resources, capabilities and knowledge. These assets are often not accessed by national or international organizations because they are either unaware of the potential or because the mechanisms that permit a dialogue with local, particularly poor, communities do not exist.

Thirdly, the paper argued that CBDRM could lead to a more appropriate and sustainable reduction of risks, given communities' role in decision making and planning. In other words, CBDRM had the potential to avoid the classic problem of white elephant projects, where disaster risk reduction measures that are apparently technically sound are rejected or abandoned by the communities they are supposed to benefit, due to a mismatch with their cultural, economic or social experience and expectations (Maskrey and Rochebrun, 1990). Already by the early 1980s, documented cases abounded of hazard-resistant but ecologically and culturally inappropriate housing, which ended up not being accepted by the local population (Davis, 1978, Monzon and Oliden, 1989). Other cases showed how investment in local infrastructure to reduce hazard, such as storm drainage or slope stabilization, without local involvement and ownership, often led to a lack of maintenance and critical disrepair.

3. From Ocho Rios to infinity: the mainstreaming of CBDRM

At the time of the 1984 Ocho Rios Conference, the bibliography on CBDRM was frugal, to say the very least. And while pioneering NGOs in Africa, Asia, the Caribbean and Latin America were implementing projects and programmes at the community level, it was unusual for these to be systematically documented and even rarer to be explicitly described as CBDRM. Also as Ian Davis remarked in his introduction to *Community Based Disaster Mitigation* (Maskrey, 1989), NGO interventions with risk-prone communities were often viewed with suspicion by governments and were not taken seriously by the scientific and technical community. In the best of cases, CBDRM was considered by most governments and international organizations as a fringe activity.

A quarter of a century later, it is difficult to find a major disaster risk management programme in the world that does not have an explicit CBDRM component or else focuses on LLDRM, building the capacity of local governments. From the fringe the approach would seem to have become mainstreamed to the point of orthodoxy. Since the 1990s, both CBDRM and LLDRM approaches have been increasingly and enthusiastically taken up by almost all the major international NGOs and have been promoted by bilateral and multilateral organizations and governments and even by organizations such as the World Bank and United Nations Development Programme (UNDP). As a practice, CBDRM has taken root in all developing regions (Sharma et al., n.d.; UNISDR and UNDP, 2007). LLDRM, in contrast, has evolved mainly in Latin America and to a lesser extent in Asia (Lavell et al., 2005, 2009).

LLDRM and CBDRM approaches are rarely mutually exclusive. Most LLDRM approaches rest on partnerships between local governments and community and civil society organizations. Similarly, the most successful CBDRM approaches are those that have managed to attract the support of local and national governments. Clearly, the definition of what is or is not local varies from context to context. As a mediator and arbitrator of different social interests and conflicts and as a key actor in environmental, territorial and sector planning and development, local governments can potentially play a major role in disaster risk reduction. However, the strength or weakness of local governments varies enormously according to a country's territorial and political–administrative structure, the level of decentralization of government responsibilities and the availability of resources.

It is difficult to identify how, why and where these approaches went from the marginal fringe of disaster risk management to become a central component of the programming of national and international organizations and NGOs. Probably a number of factors conspired together.

- Networking brought together previously isolated individuals and institutions, working from a social perspective into a dynamic movement that helped to put risk-prone communities at the centre of the disaster risk management agenda. The formation of the Network for Social Studies on Disaster Prevention in Latin America (LA RED) in 1992 (LA RED, 1992), of Duryog Nivaran (a South Asian initiative on disaster mitigation) in 1994 (Duryog Nivaran, 1995) and Peri Peri in Southern Africa in 1997, should be seen as key milestones in consolidating and deepening global awareness and understanding of CBDRM and in promoting its application.
- Research by these networks was also crucial in highlighting the hitherto ignored impact of small- and medium-scale disasters, which affect local communities but are neither documented in international disaster databases nor addressed by the international community. The assembly of a formidable body of quantitative evidence on these impacts, through the development of national disaster databases, using the DesInventar methodology developed by LA RED, was another factor that began to configure a policy environment favourable to the adoption of CBDRM approaches (LA RED, 1998).
- Publication also clearly played a key role. The aforementioned book on *Community Based Disaster Mitigation* (Maskrey, 1989) published by OXFAM at Ian Davis's behest expanded the arguments presented in the Ocho Rios paper and exposed them to a far wider audience. The book found echo in the NGO community, particularly in Asia, and encouraged many community activists to look seriously at disaster risk management for the first time. But other more widely distributed books (Anderson and Woodrow, 1989) gave further impetus to the argument that CBDRM was important. The series of books produced in Spanish by LA RED clearly had an enormous influence in Latin America, particularly *Viviendo en Riesgo* (Lavell, 1994). Publications by Duryog Nivaran in Asia and Peri Peri in Africa further reinforced the message in those regions.
- It is probably not coincidental that *Community Based Disaster Mitigation* heavily emphasized the potential role of NGOs as supporters of CBDRM. A key factor in the adoption of CBDRM has certainly been the take-up by international NGOs such as OXFAM (OXFAM, 2002), CARE (CARE, 2002), ActionAid (ActionAid, 2005) and Tearfund (Venton and Hansford, 2006) as well as by the International Federation of the Red Cross (IFRC, 2007).
- Another milestone was the incorporation of CBDRM into training and capacity development programmes. By the late 1980s, CBDRM was already being highlighted in the programmes offered by the Oxford Centre for Disaster Management in the United Kingdom. In 1997, the Asian Disaster Preparedness Centre with Duryog Nivaran

started offering the first bespoke course on CBDRM. LA RED and Peri Peri, meanwhile, also developed their own training modules and materials and began to offer workshops. By the late 1990s, therefore, a large and growing number of professionals from NGOs, local and central governments and other agencies had been exposed one way or another to CBDRM concepts, approaches and methods.

- The development of a range of tools for community risk or vulnerability assessments, which use participatory action research methods that enable people and communities to play an active role in planning, assessment, design, implementation and evaluation of disaster risk management activities also played a role. In general, such methodologies include an assessment not only of vulnerabilities but also of capacities. A large number of manuals and methodological tools have been developed in each region, many of which were surveyed and assembled in a useful compendium by the ProVention Consortium (ProVention Consortium, 2007).
- In 1994, the Yokohama Declaration, from the first World Conference on Disaster Reduction, put an official seal of blessing on these approaches by stressing the importance of focusing disaster risk reduction efforts on poor communities. Soon thereafter, bilateral and multilateral agencies began to open funding windows and develop programmes to support CBDRM. DIPECHO, for example, has promoted a large number of CBDRM projects where preparedness and response planning and early warning systems have often been complemented by local hazard mitigation schemes, such as building and strengthening dykes and stabilizing slopes. Despite a relatively modest investment (approximately €80 million worldwide over the last decade), DIPECHO has become a major resource provider for CBDRM initiatives, along with DFID, GTZ, Swiss Development Cooperation, the Italian Cooperation and many others (United Nations, 2009). The Red Cross and Red Crescent movement had also played an important role and in the last 10 years UN agencies such as UNDP and international financial organizations such as the World Bank and the Inter American Development Bank have included CBDRM in their country programming.
- Most importantly, many governments and inter-governmental regional organizations, in all regions, have now included CBDRM as a key programme component and many of the largest CBDRM initiatives in the world are now government driven. In Latin America and the Caribbean, CDERA, CEPREDENAC (Lavell, 2005) and CAPRADE (Lavell, 2009) have all embraced community-based approaches, while some of the largest-scale programmes exist in Asia (Pandey and Okazaki, n.d.), Bangladesh, India (Government of India and UNDP, 2009) and the Philippines (Victoria, n.d.), for example.

CBDRM and LLDRM approaches have been applied to: the mapping and monitoring of hazard levels, using local knowledge to develop early warning systems; hazard mitigation activities, such as the strengthening or construction of dykes, slope stabilization, the recovery of mangroves and, in urban areas, improvement of drainage; improved ecosystem management, under local responsibility, including in some cases the use of Payment for Ecosystem Services mechanisms; development of participatory land-use and development plans that incorporate disaster risk considerations; strengthening of livelihoods, through promotion of employment and measures to increase agricultural productivity, water and food security and marketing initiatives; the application of microfinance and microinsurance to increase social protection and resilience; strengthening of local governance, including the disaster risk reduction capacities of local governments; and the adoption of gender-sensitive approaches in disaster risk reduction practices. A great number of community and local development initiatives also address underlying risk factors but are not labelled CBDRM and LLDRM.

At least from a superficial glance, therefore, it would seem that if one looks forward from the 1984 Ocho Rios conference, CBDRM has been a classic success story of mainstreaming. But at the same time it is crucial to ask to what extent the promise expounded in the Ocho Rios paper has really been fulfilled.

4. Community-based disaster risk management or disaster risk management at the community level

Despite the massive take-up of CBDRM and its cousin LLDRM, there is evidence that some of what goes on under those acronyms has little to do with the approach proposed in Ocho Rios. At the same time, the experience of the last quarter-century has also highlighted limitations of that approach in practice.

Firstly, both CBDRM and LLDRM are described as processes through which disaster risk reduction issues are addressed and local capacities strengthened. In other words, an underlying rationale is that there is empowerment of and ownership by local stakeholders, either at the community or municipal level that should lead to a sustainable reduction in disaster risks over time.[1]

In practice, however, many initiatives would seem to be programmes and projects that are implemented at the community or local levels rather than with community or local ownership (United Nations, 2009). Many initiatives are promoted by NGOs and other supra-local actors and have been delivered as relatively short-term projects or programmes. These do not necessarily generate the organizational and institutional underpinning that sustainable processes would require and that was an essential feature of the approach advocated at Ocho Rios. Often, when the project or programme ends, the process also ends, indicating that local ownership and involvement may be shallower than it appears.

Secondly, in the Ocho Rios paper it was argued that CBDRM, beginning with actions at the community level, should gradually expand to catalyse disaster risk management actions at other scales. Despite the long list of applications described above, many CBDRM and LLDRM initiatives limit themselves to risk factors that can easily be addressed at the community or local level with small investments, for example, improvements in disaster preparedness and response, rather than those that require addressing more structural issues such as access to land or natural resource management.

Improving local response and early warning capacities is in itself, of course, enormously important. For example, in La Masica, Honduras, an effective locally managed early-warning system resulted in no mortalities during Hurricane Mitch, while neighbouring areas where local capacities had not been strengthened suffered hundreds of deaths. It was a similar case when the 2004 Indian Ocean tsunami struck the coast of Tamil Nadu. UNDP's strengthening of local capacities in the village of Samiyar Pettai led to a greatly reduced impact (United Nations, 2009).

In disaster preparedness and response, the required investments are relatively small and the benefits immediately visible, at least when the next disaster strikes. At the same time, local preparedness and response capacities can generally be strengthened without addressing underlying risk drivers such as land tenure and access to resources, which generally lie outside community and local government control. Success in expanding beyond the community level to engage those stakeholders such as government, which do regulate land tenure, for example, has been far more limited, precisely because so many of the factors that need to be addressed are beyond the capacities of local stakeholders to address.

A sobering reality check is that in parallel with the massive take-up of CBDRM approaches, disaster risk has continued to increase, even without taking into account potential increases in hazard severity and frequency due to climate change (United Nations, 2009). In terms of one of the underlying risk drivers, it is currently estimated that more than 1 billion people live in squatter and informal settlements in developing countries (Dodman et al., 2008), many of these in hazard-prone areas. This number is growing by an estimated 25 million people per year. These data tend to corroborate the hypothesis that while CBDRM and LLDRM approaches may have led to significant improvements in disaster preparedness and response, they have had little impact on underlying risk drivers, such as the lack of provision of land for low-income households in safe areas. CBDRM and LLDRM, therefore, may have had only a tangential impact on disaster risk accumulation.

Thirdly, experience has revealed that the approach advocated in Ocho Rios in 1984 also needs to be critiqued.

While influenced by the prevailing political context in Peru and Latin America at the time, the approach was based on a polarized vision of the relationships between civil society and government. The paper stated that

> although governments in many contexts intervene to implement mitigation measures at the community level, this usually has the object of defusing rather than enabling community awareness and organisation ... when governments are openly repressive, of course, community based activities are considered as subversive and treated as such.

This was not just ideological speculation but reflected many of the realities and dangers faced by NGOs working to empower communities in Latin America at the time, and not least by community leaders themselves. But subsequent experience has shown that many of the best and most sustainable CBDRM programmes are those that have been promoted by governments, while some NGO interventions have been limited precisely due to an unwillingness to engage government. The last 25 years have highlighted the existence of many shades of grey in the stark black and white political landscape portrayed in the Ocho Rios paper.

Similarly, the approach oversimplified the dynamics of the dialectic relationships between communities, governments and NGOs across different scales, making the expansive nature of CBDRM, as communities engage other stakeholders and address increasingly deeper vulnerabilities at increasing spatial scales, seem both inevitable as well as a one-way street. Experience has shown that all such processes are far more complex, with ebbs, flows and eddies as both organizations themselves and the external environment change and evolve. Every CBDRM experience produces ripples and waves that expand outwards but not necessarily in a linear one-directional way.

5. CBDRM: the future

Local areas exposed to the same hazard manifest very different patterns of risk. While disaster risk is influenced by broader national and global factors such as the quality of governance or climate change, ultimately it is shaped at the local level. Risk and its manifestations are always experienced, not in the abstract, but in a place and time. Communities, municipal governments, enterprises and other local actors use and transform territory, natural resources, the built environment and other assets, in the context of processes at other scales. In doing so they configure locally specific patterns of hazard, exposure, vulnerability and resilience, generating what could be described as a social territory of risk.

New data on local-level disaster impacts, compiled since the mid-1990s, provide an empirical foundation to assert the importance of place, which was not available when the Ocho Rios paper was written. Such data have highlighted two important patterns. Firstly, they show how even major disasters, associated with extreme hazard events, have highly heterogeneous impacts at the local level, reflecting widely varying patterns of risk accumulation, associated in turn with highly idiosyncratic social territories of risk. For example, Hurricane Mitch affected large areas of Honduras in Central America at the end of October 1998 and was one of the most destructive Atlantic hurricanes ever experienced. Nevertheless, despite the commonly held view that Mitch has destroyed the entire country, mortality patterns were highly heterogeneous, indicating very idiosyncratic local risk patterns (see Figures 1 and 2).

Secondly, the data show that disasters such as those associated with Hurricane Mitch occur only infrequently. Frequently occurring, highly localized off the radar screen disaster events occur in their thousands every year and are responsible for a very significant proportion of damage to housing, crops, livestock and local infrastructure such as roads, schools, water, sanitation and so on. From a sample of 12 Asian and Latin American countries, 99.3 per cent of loss reports from local administrative areas were associated with this kind of event (approximately 126,000 reports since 1980 in the 12 countries) and were associated with 51 per cent of housing damage. Therefore, except when extreme hazards make the headlines, disaster data show that disaster risk management is fundamentally a local affair (United Nations, 2009).

If risk is configured and disaster impacts are experienced quite uniquely in given places and times, then disaster risk management has to be embedded and grounded in each social territory of risk. As such, community-based and local-level approaches, which engage local stakeholders, civil society organizations and municipal governments, will continue to be critical for reducing disaster risk. As highlighted above, the experience of the last 25 years seems to show that CBDRM and LLDRM are more effective in achieving

FIGURE 1 Hurricane Mitch, 27 October 1998.
Source: NOAA, 1998.

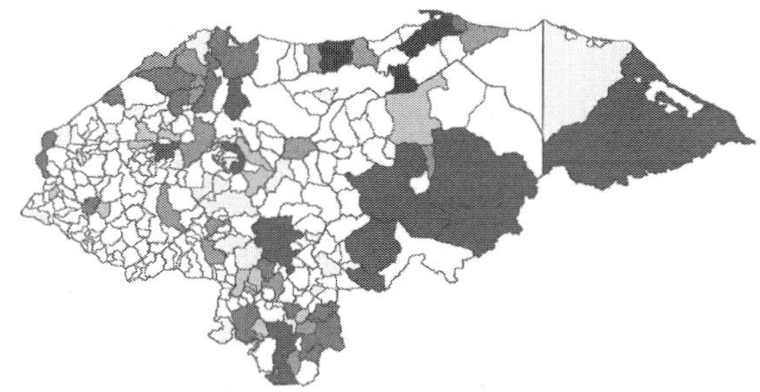

FIGURE 2 Mortality by municipality, Hurricane Mitch.
Source: COPECO, 1998.

improvements in preparedness and response than addressing risk factors that are linked to issues such as land ownership or that require actions at higher spatial scales. However, in many ways the tenets laid out in Ocho Rios 25 years ago are still valid and in many ways vitally contemporary. There is evidence from the most successful CBDRM and LLDRM programmes that not only can the potential outlined in Ocho Rios be fulfilled, but also these approaches are more relevant than ever in a context where disaster risk continues to increase in developing countries and where climate change will aggravate and magnify pre-existing risks.

Recent publications, particularly from Latin America (Lavell, 2009), indicate that a growing number of CBDRM and LLDRM initiatives are now moving beyond disaster preparedness and response to address the vulnerability of livelihoods, the decline of ecosystems, the lack of social protection, unsafe housing, the improvement of governance and other underlying risk factors. Others aim to factor disaster risk considerations into local land-use and development planning, for example.

The most successful programmes of all are those that – while community or locally based – have developed broader partnerships with governments and other supra-local stakeholders (Government of India and UNDP, 2009). As stressed above, many of the underlying risk drivers cannot be addressed by community organizations or local governments on their own. Partnerships with national agencies permit scaling-up of initiatives to go beyond individual communities and localities to address problems that affect wider areas, such as watersheds and coastlines. They enable the investment of resources that are unavailable locally and increase continuity and sustainability as initiatives move from stand-alone projects and programmes to longer-term processes.

Many of these more successful initiatives would appear to have been catalysed by decentralization processes, in which more competent and better resourced local governments are able to play a more active role in addressing disaster risk. Those cases where sustainable local processes have emerged are where national governments have decentralized both responsibilities and resources to the local level, as for example in cities in Colombia such as Manizales. In Bangladesh and Cuba, success in disaster preparedness and response, leading to a real and drastic reduction in mortality due to tropical cyclones, builds on solid local organization, but in both cases it has received sustained support from the national level.

For example, LLDRM has been most successful where decentralization processes have provided local governments with the resources and capacities to fulfil their disaster risk reduction responsibilities, such as in Brazil or Colombia. In such contexts, local governments often become promoters of CBDRM processes. Where local governments are weak and disempowered it is unlikely that LLDRM will take root as the technical capacities, political authority or financial resources required are generally not present.

There is now evidence from a host of initiatives both in disaster risk management and other areas such as sanitation, urban upgrading and strengthening rural livelihoods (United Nations, 2009) that shows that if such partnerships can develop, the advantages of cost-effectiveness and appropriation highlighted in the Ocho Rios paper can be fulfilled.

An enabling environment for such partnerships to be nurtured and grown would seem to have elements that include competent and accountable local government, a strong civil society that can hold the government to account and a supportive policy framework that has incorporated disaster risk management into national development plans and budgets.

To conclude, CBDRM and LLDRM approaches remain valid and more necessary than ever if the goals of international frameworks such as the Hyogo Framework of Action have to be achieved. However, their potential will only be fulfilled when they are nurtured in the context of balanced partnerships between government and civil society, based not only on local participation and ownership but on political and economic support from national institutions. Global political commitment is required to address global risk drivers, such as climate change. National political commitment is required to address issues such as land ownership and distribution and the management of water resources. But effective risk management will have to be addressed in local scenarios. As such the future will depend on moving beyond the polarized vision of top-down and bottom-up processes profiled in the Ocho Rios paper, towards partnerships that make linkages and mediate between the different scales of action. CCDRM and LLDRM on the one hand and supportive national programmes and policy on the other hand are mutually supportive. As such, the future of disaster risk management will depend largely on political processes capable of developing constructive partnerships between the national and local domains and between government and civil society.

Note

1. Note, for example, the following definition: 'the process of disaster risk management in which communities at risk are actively engaged in the identification, analysis, treatment, monitoring, and evaluation of disaster risks in order to reduce their vulnerabilities and enhance their capacities. This means that people are at the centre of decision making and implementation. The involvement of the most vulnerable is paramount and the support of the least vulnerable necessary. Local and national government are involved and supportive' (Abarquez and Murshed, ADPC, 2004).

References

Abarquez, I. and Murshed, Z., 2007. *Community Based Disaster Risk Management: A Field Practioner's Handbook*. ADPC, Bangkok.

ActionAid, 2005. *Participatory Vulnerability Analysis: A Step by Step Guide for Field Staff*. ActionAid, London.

Anderson, M. and Woodrow, P., 1989. *Rising from the Ashes: Development Strategies in Times of Disaster*. Lynne Rienner Publishers, Boulder.

Bommer, J., 1985. The politics of disaster: Nicaragua. *Disasters*, 9(4).

CARE, 2002. *Household Livelihood Security Assessments: A Toolkit for Practitioners*. CARE USA, Atlanta.

COPECO, 1998. *DesInventar: Hurricane Mitch*. LA RED, Colombia.

Davis, I., 1978. *Shelter after Disaster*. Oxford Polytechnic Press, Oxford.

Dodman, D., Hardoy, J. and Satterthwaite, D., 2008. *Urban Development and Intensive and Extensive Risk*. IIED, London.

Egas, R., 1985. Ecuador: inundaciones 1982–1983 en la Cuenca baja del Guayas: Procesos de organizacion de los campesinos para hacer frente al desastre, Sri Lanka. *Desastres Naturales y Sociedad en America Latina*, in J. Hardoy, et al. (eds), Grupo Editora Latinoamericana.

Freire, P., 1970. *Pedagogy of the Oppressed*. Continuum Publications, New York.

Government of India and UNDP, 2009. *Good Practices in Community Based Disaster Risk Management*. United Nations Development Programme, New Delhi.

IFRC, 2007. *How to do A Vulnerability and Capacities Assessment*. International Federation of the Red Cross and Red Crescent Societies, Geneva.

LA RED, 1992. *Research Agenda and Constitution*. ITDG Publications, Lima.

LA RED, 1998. *DesInventar: Disaster Information System*. Corporacion OSSO, Cali.

Lavell, A. (compiler), 1994, *Viviendo en riesgo: comunidades vulnerables y prevencion de desastres en America Latina, FLACSO, CEPREDENAC, LARED*. Tercer Mundo Editores, Bogota.

Lavell, A. and Lavell, C., 2009. *Local Disaster Risk Reduction: Lessons from the Andes*. PREDECAN, Lima.

Lavell, A., Mansilla, E. and Smith, D., 2005. *Local Level Risk Management: Concept and Practices*. CEPREDENAC and UNDP, Ecuador.

Leslie, J., 1986. A building education programme in North Yemen. *Disasters*, 10(3).

Maskrey, A., 1984, Community based disaster mitigation. Proceedings of the International Conference on Disaster Mitigation Program Implementation, Ocho Rios, Virginia Polytechnic Institute, 12–16 November.

Maskrey, A., 1989. *Community Based Disaster Mitigation*. OXFAM Publications, UK.

Maskrey, A. and Rochebrun, G., 1990. *Y Si Dios hizo la Noche sin Luz*. ITDG Publications, Lima.

Maskrey, A. and Romero, G., 1985. *Urbanizacion y Vulnerabilidad Sísmica en Lima Metropolitana*. PREDES, Lima.

Monzon, F. de. M. and Oliden, J., 1989. *Vivienda Popular Y Tecnología*. ITDG Publications, Lima.

Nivaran, D., 1995. *Disasters and Vulnerability in South Asia: A Programme of Work for Duryog Nivaran*. Intermediate Technology Development Group (ITDG), Sri Lanka.

NOAA, 1998. *Hurricane Mitch Special Coverage*. National Environmental Satellite, Data and Information Service (NESDIS), Boulder, CO.

OXFAM, 2002. *Participatory Capacities and Vulnerabilities Assessment: Finding the Link Between Disasters and Development*. OXFAM, Philippines.

Pandey, B. and Okazaki, K., n.d. *Community Based Disaster Management: Empowering Communities to Cope with Disaster Risk*. UNCRD, Japan.

ProVention Consortium, 2007. *Community Risk Assessment Toolkit*. ProVention Consortium, Geneva.

Sen, A. K., 2000. *Development as Freedom*. Alfred A. Knopf, New York.

Sharma, A., Gupta, M. and Kaushik, R., n.d. *Community-based Disaster Management: Lessons from Asia*. SEEDS, India.

UNISDR, 2009. *Global Assessment Report on Disaster Risk Reduction*. Global Network of NGOs for Disaster Risk Reduction, Geneva.

UNISDR, UNDP, 2007. *Building Disaster Resilient Communities: Good Practices and Lessons Learned*. Global Network of NGOs for Disaster Risk Reduction, Geneva.

Venton, P. and Hansford, B., 2006. *Reducing Risk of Disasters in our Communities*. TearFund, London.

Victoria, L. P., n.d., Community based disaster management in the Philippines: making a difference in people's lives, manuscript (grey literature).

Wilches-Chaux, G., 1989. *Desastres, Ecologismo y Formacion Profesional*. SENA, Colombia.

Wisner, B., 1979. Flood prevention and mitigation in the Peoples Republic of Mozambique. *Disasters*, 3(3).

United Nations, 2009. *2009 ISDR Global Assessment Report on Disaster Risk Reduction*. UNISDR, Geneva.

ENVIRONMENTAL HAZARDS
Human and Policy Dimensions

Disaster risk reduction and the evolution of physical development regulation

FREDRICK KRIMGOLD*

The World Institute for Disaster Risk Management, Advanced Research Institute, 4300 Wilson Blvd., Suite 750, Arlington, VA 22203, USA

In 1984 the Mitigation Implementation Conference recognized the need for effective management of urban development in terms of building standards and land-use zonation. In the intervening 25 years, significant progress has been made in the development of building codes and hazard mapping for disaster risk reduction; however, the failure to promulgate, implement and enforce regulation of development has resulted in the dramatic expansion of exposure and risk. It is now necessary to re-assess regulatory policy and practice. New approaches to regulatory compliance must be explored, including participatory regulation and reliance on cultural and religious bases for individual commitment to community safety.

Keywords: building regulation; building standards; disaster mitigation; land-use management; loss reduction; natural hazards; risk management

1. Ocho Rios 1984: mitigation implementation

In the 1984 Ocho Rios Conference, the topic was Natural Hazard Program Implementation. At that time, the primary natural hazards under consideration were flood, hurricane and earthquake. This focus reflected the architectural and engineering background of many of the participants. The primary mitigation measures under consideration were land-use management and building standards. To the extent that areas of exposure for flood, hurricane and earthquake could be mapped in terms of frequency and intensity of occurrence, it should be possible to avoid development in hazardous areas or to require construction capable of resisting expected loads.

While programmes of riverine and coastal flood mapping and seismic risk mapping had been initiated in developed countries, most developing countries lacked any detailed hazard mapping or risk zonation. Similarly, few developing countries had officially promulgated building codes for construction in areas subjected to flood, wind or earthquake.

In 1984 there was a serious call for the rapid expansion of hazard mapping at a scale that would provide meaningful guidance to urban planners and local builders. There was also a call for the expansion of buildings regulations to cover indigenous construction types and critical infrastructure.

2. Regulatory approaches to land-use management and buildings standards for natural disaster risk reduction

In the intervening years, great strides have been made in the methods of hazard mapping and building design. Importantly, most settled areas

■ **Email:* Krimgold@vt.edu

doi:10.3763/ehaz.2011.0006 ISSN: 1747-7891 (print), 1878-0059 (online) www.earthscan.co.uk/journals/ehaz

of the planet are now covered at least by preliminary hazard mapping delineating exposure to flood, coastal flooding, landslide and extreme ground shaking (cite: GERM).

New methods of remote sensing have made possible rapid and relatively inexpensive mapping to terrain and the application of geographic information systems has made hazard microzonation possible (cite: Turkey study) (COGSS, 2010).

Also, during the past 25 years, building standards have been developed for virtually all types of engineered and 'non-engineered' construction. National and international building standards have been consistently revised and improved following every major event. Recent events in those countries where building codes are rigorously applied have shown a dramatic improvement in the performance of new conforming structures (cite: Chile).

In 2004, the International Association of Earthquake Engineering published the national seismic design codes of 54 countries (IAEE, 2004). This represents significant progress over the quality and distribution of seismic design standards 20 years earlier. Beyond these formal seismic design standards, a considerable number of informal guides for designers and builders of simple structures have been developed in local languages with rich illustration (cite UNESCO and School Guidelines).

The International Association of Earthquake Engineering has published a range of documents providing guidance on the development of seismic design provision for model codes:

- Regulations for Seismic Design – A World List, 2004/1220 pages
- Regulations for Seismic Design – A World List, 1996/1280 pages
- Basic Concepts of Seismic Codes, Vol. I., 1980
- Basic Concepts of Seismic Codes, Vol. II., 1981
- Guidelines for Earthquake Resistant Non-Engineered.

Similar guidance has been provided for wind and flood design standards from international sources.

3. Caribbean example: Caribbean Uniform Building Code

In the Caribbean, the Caribbean Uniform Building Code (CUBiC) stands as a notable example of building standards development.

CUBiC has provided the basis for various national codes with particular guidance for wind and seismic design. The Organization of Eastern Caribbean States has developed a code based on CUBiC. CUBiC is generally comparable to US and Canadian model codes (Gibbs, 2003a).

The official promulgation of building codes based on CUBiC is not uniform across the Caribbean. This is representative of the state of building regulation in many disaster-prone developing countries. While appropriate model codes that deal with relevant hazards have been developed, they are not yet officially promulgated and lack the force of law. Often the political process is heavily influenced by building owner and developer interests (Gibbs, 2003b).

3.1. Caribbean example of the importance of regulatory implementation: Saint Martin/Sint Maarten

In September 1995 Hurricane Luis passed to the northeast of the island of Saint Martin/Sint Maarten. The event offered a unique opportunity to compare the consequences of differing building regulatory policies. Saint Martin is French and Sint Maarten is Dutch. As documented by engineer Tony Gibbs, the amount of damage in Sint Maarten was significantly more than in Saint Martin, although the French side of the island was more exposed to the centre of Luis than was the Dutch side. There is no clear evidence that Sint Maarten experienced higher wind speeds than the French side. The amount of damage caused was significant. In the case of Dutch Sint Maarten, the damage was catastrophic. Direct losses were equivalent to the gross domestic product (GDP) and indirect losses added a similar amount, for a total loss of the order of twice the GDP. This level of damage was not experienced on the French side. On-site observers during and immediately after the event confirmed that the differences in levels of damage between Saint Martin and Sint Maarten were stark.

4. The differing regulatory regimes

Engineers and builders who had worked on both sides of the island were consulted on the dramatic contrast in damage experienced on the two sides of the island. Mr Ronald Daal of Independent Consulting Engineers (ICE), with headquarters on the Dutch side, indicated that there were significant differences in the regulatory regimes on the two sides of the border. ICE maintains offices in both territories. On the Dutch side, the buildings are designed in accordance with a variety of standards, including those of the Netherlands. The checking authority is the government Public Works Department, although this task is occasionally contracted out to private firms. On the French side, construction must comply with the French 'norms' and the design and construction are checked by bureaux de contrôle. In Mr Daal's words, on the French side 'you have to do it right'. The Contractors Association in Sint Maarten, which includes architects, engineers, builders and government officials, has considered various issues related to the Luis experience in Sint Maarten and the way forward for the building industry. The contrast with French Saint Martin was examined. The differences outlined by those familiar with construction on both sides of the border included: 'better attention to conceptual design on the French side; greater consistency and uniformity of standards of design for earthquakes and hurricanes on the French side; the involvement of bureaux de contrôle on the French side'.

The example of Saint Martin/Sint Maarten may present as close to a 'laboratory' test comparison of differing approaches to building regulation and their implications for damage control. The case also points to the critical importance of the mechanisms of code implementation and administration. The mechanics of the building regulatory process have not received adequate attention from the building professions, the building industry, local and national governments, or the international donor community.

Over the past approximately 30 years, international development has been powerfully influenced by emphasis on free market economic philosophy including privatization and deregulation. In this climate, very little attention has been focused on formal regulation of urban development, building standards or land-use management. During a period of rapid urbanization over the past 30–40 years, disaster risk has expanded dramatically due to unregulated growth of the built environment.

Building standards and hazard mapping now available provide the basis for effective regulation to reduce disaster risk. There is abundant evidence of the effectiveness of adequate building regulation and land-use management in limiting disaster losses. The stark contrast of earthquake losses in Haiti (January 2010) and Chile (February 2010) is directly related to quality of construction and the absence of an effective building regulatory function in Haiti.

5. Components of a regulatory system

Improvement of quality of construction is a complex issue requiring significant public/private collaboration and recognition of common purpose. Improvement of building quality is not simply a question of enforcement. Quality improvement must address the barriers of poverty, ignorance and corruption. In most cases, better quality of construction involves at least some increment of increased cost. The regulatory process itself typically adds 1 per cent to construction cost. Improved quality of building materials (appropriate amounts and quality of cement, aggregate and rebar, for example). The added cost of safe construction is estimated to be between 5 and 10 per cent of construction cost. Further, the cost of education, training and retention of an appropriately trained design and construction workforce will contribute to increased initial construction cost. These small increments of increased initial construction cost are readily balanced by the reduction of future disaster loss of life and property. Even in the developed world the most effective and reliable way to ensure the inclusion of adequate risk reduction measures in initial construction is by way of regulation. Regulatory standards represent a societal judgement of acceptable risk to which all members of society must adhere. Elevation of this level of acceptable risk is an essential component of social and economic development.

In the aftermath of disaster, there are typically reports of corruption by builders or building regulators. Experience has shown in the case of the most recent earthquake disasters that failures are most often the result of ignorance rather than criminal intent.

An essential component of building quality improvement is improved training of designers and builders. Engineers and architects must be trained in the specifics of design for disaster risk reduction. In general, there is a shortage of trained professionals in the building professions.

In many disaster-prone developing countries, there are no adequate instructional programmes for engineers and architects or builders on seismic-, wind- or flood-resistant design.

Indigenous or non-engineered construction has been a major source of building failure in the past. Adobe and unreinforced masonry are known to perform poorly under dynamic loads; however, in recent earthquakes catastrophic building failures have resulted from the inappropriate use of reinforced concrete, a modern material and construction technology. A major problem of reinforced concrete construction is that it appears to be relatively easy and it is, in fact, relatively easy to build a reinforced building to resist gravity loads. Unfortunately, reinforced concrete construction to resist lateral loads requires considerably more sophisticated design, material management and construction. Overcoming this 'invisible' difference between safe and unsafe reinforced concrete construction provides the imperative for formal training of designers, builders and regulators.

The educational component of building quality improvement must include formal training and certification of building designers. The cadre of qualified designers must be developed to meet the demands of the building industry. In many countries, these categories of specialists are in short supply. In this case, those qualified professionals must be used efficiently to provide standardized designs and to supervise junior designers.

It is particularly important that natural hazard-resistant design be introduced into the curriculum of polytechnic and vocational training programmes for construction site supervisors who will have direct responsibility for construction practice. Vocational education programmes must be provided for construction labourers as well. Labourers tasked with mixing concrete must understand the importance of material quality and ratios of concrete mixing. Barbenders must understand the importance of rebar placement and connections (Gulkan, 2001). Numerous training programmes have been developed for 'masons' in earthquake-prone areas. These are very important; however, special attention must now be paid to reinforced concrete construction because it is so prevalent and so potentially dangerous.

The improved quality of the building design and construction workforce will require coordinated action by universities, polytechnics, vocational schools, professional societies, labour unions, the building industry and the building materials industry. The quality of educational and training efforts must be tested and certified to ensure that ignorance has, in fact, been overcome.

The assurance of safe construction requires a functioning regulatory process.

The feasibility of the regulatory process is dependent on:

- The general societal appreciation and acceptance of the need for a standard of safety.
- Private sector acceptance and support for the principle of improved building safety and the incremental cost of safety.

- Political and public sector recognition of the necessity of improved building safety.
- Training for a cadre of adequately trained architects and engineers and construction supervisors.
- Certification programmes for design and construction professionals, materials testing laboratories and supervisory design firms.
- Community support for the enforcement of building standards, including demolition of non-conforming construction.

6. Barriers to effective building regulation

In most developing countries, the technical resources in terms of trained personnel are limited. The building regulatory process requires competent professionals for the functions of plan review and site inspection. Typically, local government salaries are not adequate to attract or hold qualified personnel, particularly in rural areas.

The problem of low compensation also makes building regulators vulnerable to corrupt practices. Building owners and developers may find it less expensive to bribe a regulatory official than to conform to relevant building standards.

Local government is often dominated by building owners and developers who are not predisposed to limiting the prerogative of investors and builders. This circumstance may weaken the capacity for regulatory enforcement.

7. Innovation in building quality improvement

Model building codes and land-use zoning standards have been developed to provide the basis for development regulation in most parts of the world. It is now necessary that these model provisions be adapted to local conditions and values and be promulgated as law. Local adaptation of model standards should include adaptation to the specific building culture and incorporation of local standards of community safety. In order to conform to these promulgated standards, it is necessary to create the capability on the part of the building professions and the building industry to meet the requirements of the building law.

It is not reasonable to assume that the building regulatory systems in developing countries will be initiated as copies of building regulatory practice in the developed world. Human, material and financial resource constraints may make it impossible to copy the local public bureaucratic model of the developed world. Several new approaches may be explored to introduce the building regulatory function (Imamura, 2010). Participatory regulation has been attempted at the community scale.

In the initial stage of introducing a building regulatory mechanism, it may be necessary to emphasize compliance assistance rather than police action for enforcement. The regulatory process must provide a comprehensive approach to building safety. A critical function of the regulatory process is to educate designers and builders on appropriate practice. This would suggest that emphasis be placed on educating and correcting errors in design and construction practice.

As a part of community involvement is disaster management, the population has been offered a basic understanding of safe building practices and observable indicators of good or bad practice. This knowledge allows the public to exercise a monitoring function and to exercise their interest in community safety. It also mobilizes the force of popular judgement and reputation in favour of safe construction. Name and shame becomes an antidote to poor construction.

Building owners and developers should be mobilized as principal advocates of building quality improvement. The business case for safe building must be developed and supported by public policy. Lenders and insurers should be enlisted in support of investment in risk reduction and safe construction.

Finally, contribution to community safety is a moral obligation. Individual building owners, designers and builders should be held morally accountable for implications of their actions for the safety of the community. Community moral authorities and religious leaders should be enlisted in support of building quality improvement and disaster risk reduction.

The tragedy of the Haiti earthquake bears stark witness to the failure of building regulation and development control. Recent surveys of regulatory effectiveness in the Caribbean indicate that this problem is not limited to Haiti. The regulatory initiative is imperative. It remains the key missing link in natural disaster mitigation.

References

COGSS, 2010. *Draft Guidance on School Construction*, Geneva.

Gibbs, T., 2003a. *Comparison of Building 'Codes' and Practices in Use in the Caribbean*. Pan American Health Organization, May.

Gibbs, T., 2003b. *The Use of Check Consultants in Design and Construction*. PAHO Leaders Course, Ocho Rios, Jamaica, February.

Gulkan, P., 2001. *Rebuilding the Sea of Marmara Region: Recent Structural Revisions in Turkey to Mitigate Disasters. Challenges in Managing Catastrophic Risks: Lessons for the US and Emerging Economies*. World Bank, Washington, DC, January.

Imamura, T., 2010. *GTFBC for Haiti*, UNESCO, Paris.

International Association of Earthquake Engineering (IAEE), 2004. *Regulations for Seismic Design – A World List for 2004*. Tokyo.

ENVIRONMENTAL HAZARDS
Human and Policy Dimensions

Establishing public accountability, speaking truth to power and inducing political will for disaster risk reduction: 'Ocho Rios + 25'

RICHARD STUART OLSON[1,2,*], JUAN PABLO SARMIENTO[1] AND GABRIELA HOBERMAN[1,2]

[1]Latin American and Caribbean Center, Florida International University, Miami, FL, USA
[2]Department of Politics and International Relations, Florida International University, Miami, FL 33199, USA

The problem is that disaster risk reduction (DRR) is still more a zone of political darkness than of light. Therefore, the purpose of our contribution to this theme issue is to illuminate at least some of that zone of political darkness and to suggest a more proactive, but for many career professionals a personally problematic, way forward for DRR. The career problematic derives from the fact that the way forward would involve moving from applauding the emergence of elected and/or appointed champions (from mayors to cabinet ministers to presidents/prime ministers) for *action* to raising the political costs of *inaction* by leaders not otherwise disposed. The essential element of this way forward would be greatly enhanced public accountability. We will draw preliminary accountability lessons from the Haiti earthquake catastrophe and the Chile earthquake disaster 2010 events towards the conclusion of this paper as we outline a possible future paradigm.

Keywords: paradigm; politics of disaster; public accountability; vulnerability reduction

1. The 'politics of disaster' in 25-year perspective

At Ocho Rios, Jamaica, in the meetings of 1984, and despite relatively recent major disasters in Latin America and the Caribbean that had major political effects (including earthquakes in Managua in 1972 and Guatemala in 1976, and Hurricane David in Jamaica in 1979), an explicit focus on what has come to be called the 'politics of disaster' was largely missing.[1] Metaphorically speaking, it was a blip on the 1984 radar screen – present but low and not tracked. Indeed, the best one could hope for in most professional meetings of that time were discussions of disaster or emergency management 'policies', as if those were somehow immaculately conceived and did not require political processing.

Much has changed in 25 years. Now, the political aspects of disasters, and especially the political leadership and commitment ('political will') required for truly effective disaster risk reduction (DRR), loom large as concerns and at times even dominate the professional radar screen. The problem is that DRR is still more a zone of political darkness than of light. Therefore, the purpose of our contribution to this theme issue is to illuminate at least some of that zone of political darkness and to suggest a more proactive, but for many career professionals a personally problematic, way forward for DRR. The career problematic derives from the fact that the way forward would involve moving from applauding the emergence of elected and/or appointed champions (from mayors to cabinet ministers to presidents/prime ministers) for *action* to raising the political costs of *inaction* by leaders not otherwise disposed. The essential element of this way forward would be greatly enhanced public accountability.

*Corresponding author. ***E-mail:* olsonr@fiu.edu**

doi:10.3763/ehaz.2011.0007 ISSN: 1747-7891 (print), 1878-0059 (online) www.earthscan.co.uk/journals/ehaz

The seriousness of the issues surrounding DRR, and the stakes involved for communities and countries with literally millions of people at risk have been recently highlighted by a pair of 2010 events: the Haiti earthquake catastrophe and the Chile earthquake disaster. We will draw preliminary accountability lessons from these hazard events towards the conclusion of this paper as we outline a possible future paradigm.[2]

2. DRR: two paradigms, one old, one new

The dominant traditional professional paradigm for engaging in DRR has been, and in fact remains, 'meeting, talking, writing – and hoping' (MTWH). The core of this paradigm has been for scholars, professional practitioners and technical experts interested in DRR (or 'mitigation' in earlier times) to meet in various venues to share information and experiences, write up and exchange the results, and wait for a political and/or policy champion to emerge in a country or community and then support that champion.

The MTWH paradigm was already well established in 1984 at the Ocho Rios meeting and had not changed much by the 2009 Montego Bay meetings. The problem is that, meanwhile, vulnerabilities have increased massively in most Latin American and Caribbean (LAC) countries because of population growth and, especially, unplanned and 'uncoded' urbanization. As we will develop further below, the most recent example is obviously Port-au-Prince, Haiti, with the 12 January 2010 earthquake that killed at least 230,000 people out of an estimated urban population of between 2.5 and 3 million. To be fair, a more optimistic Latin American counter-example would be with the Chile earthquake and tsunami of 27 February 2010, which together killed 521 people.[3] The problem is that most hazard-prone countries in the LAC region fall closer to the Haiti side of the continuum than to the Chile side.

A very different and much more proactive paradigm for DRR is thus indicated, which we are calling here the 'inducing political will' (IPW) paradigm. Under the IPW paradigm, political and economic elite decision making with risk implications for a community, region or nation would be subject to public monitoring, evaluation and reporting. The idea underlying this paradigm shift is to literally *force* DRR onto political and policy agendas at all relevant levels and across all relevant sectors and provide a combination of spotlight and microscope on development/redevelopment proposals or actions that have hazard – and therefore risk – implications. The key to this new 'IPW' paradigm for DRR lies in establishing unprecedented levels of accountability, a concept that itself requires substantial explanation.

3. Accountability: definition and application

'Accountability' as a concept has become iconic and needs to be rescued, even from its friends and advocates, as Dubnick (2002) has so cogently noted. Indeed, accountability has become 'one of those golden concepts that no one can be against' (Bovens, 2007, p. 449). The problem with being invoked as a panacea, however, is that 'the concept has become less useful for analytic purposes and today resembles a dustbin filled with good intentions, loosely defined concepts, and vague images of good governance' (Bovens, 2007, p. 449; see also Bovens, 2005, 2006). That is, accountability has become too many things to too many people, the net result of which is overuse on one hand and a loss of utility on the other. The solution, we would argue, is to define it relatively closely and apply it to particular policy domains.

Working from Paul (1992), Manin et al. (1999), Mulgan (2000), Behn (2001) and Moreno et al. (2003), John Ackerman offers this definition of accountability: 'a pro-active process by which public officials inform about and justify their plans of action, their behavior, and results and are sanctioned [positively or negatively] accordingly' (Ackerman, 2005, p. 6). This definition has two principal advantages: (1) it is forward-looking and thereby avoids the trap of examining only ex-post actions, which gives accountability a potential role in policy agenda-setting, and (2) it recognizes that rewards, not just punishments, may properly be associated with accountability.

Informed by a partially overlapping but then subsequent literature, including Mulgan (2000, 2003), Strøm (2000, 2003), O'Connell (2005) and Koppell (2005), Mark Bovens has critically observed that accountability now 'often serves as a conceptual umbrella that covers various other distinct concepts, such as transparency, equity, democracy, efficiency, responsiveness, responsibility, and integrity' (Bovens, 2007, p. 449). Eschewing such a broad approach, Bovens posits a more succinct and much more operationally useful definition:

> [Accountability is] a relationship between an actor and a forum, in which the actor has an obligation to explain and to justify his or her conduct, the forum can pose questions and pass judgement, and the actor may face consequences (Bovens, 2007, p. 450).

Developing an analytic framework for accountability, Bovens (2007, pp. 454–455) then explains that an 'actor' may be an individual or an organization as agent, the 'forum' serves as the principal in a principal–agent relationship and the 'obligation' may be either formal or informal, as may be the consequences. Bovens then identifies four core questions that must be answered for accountability to have analytic meaning in any specific domain and context: (1) What is the forum? (2) Who is the proper actor/agent to render account? (3) What is to be explained and justified? and (4) Why should the actor/agent feel obligated to participate?

The problems with the Bovens definition of accountability are that (1) it is not adequately forward-looking and therefore seems to discount policy agenda-setting or at least agenda-influencing possibilities, (2) it makes the forum rather passive and (3) it emphasizes at least implicitly negative consequences. One solution, which we adopt here, is to synthesize the strengths and insights of the Ackerman and Bovens definitions to yield the following:

> Accountability is a relationship between an actor and a forum, in which (a) the actor has an obligation to explain and justify his or her plans of action and/or conduct, (b) the forum may pose questions, require more information, solicit other views, and pass judgement, and (c) the actor may see positive or negative formal and/or informal consequences as a result (see Figure 1).

This accountability definition and Figure 1 schematic, combined with the four core analytic questions from Bovens, will allow us to identify and confront issues surrounding accountability when attempting to apply it to specific policy domains and contexts, such as DRR. As noted above, this 'coming to grips' with real-world DRR-accountability is crucial if the shift to an IPW paradigm is to be even remotely viable.

Before turning to the complex of issues surrounding DRR-accountability, however, we have to delve into another definitional problem, the one surrounding the 'disaster' concept itself. It is a definitional thicket all its own.

4. Disasters and 'DRR-accountability'

In a famous aphorism from the US Supreme Court about limits to freedom of speech, Justice Potter Stewart once said that he was unable to define hard-core pornography, but he 'knew it when he saw it'. The same aphorism could easily be applied to the term disaster, where the academic community that focuses on disasters has spawned a literature of its own, including two major edited books (Quarantelli, 1998; Perry and Quarantelli, 2004) where 23 international contributors essentially came to 23 different definitional conclusions.

The academic debate continues and shows no sign of consensus, but we would like to single out for later use here an earlier definition from Anthony Oliver-Smith and Susanna M. Hoffman

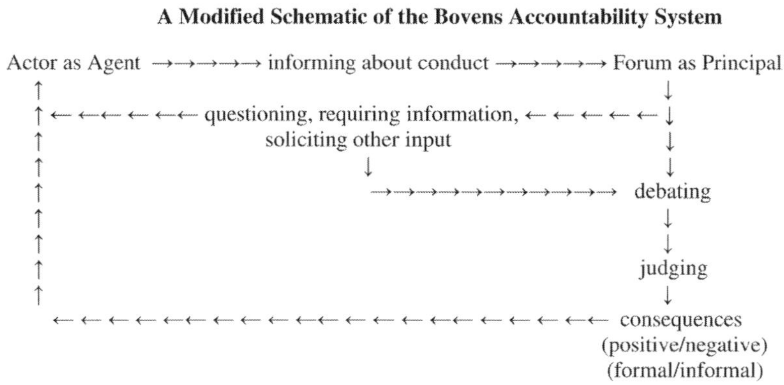

FIGURE 1 A modified schematic of the Bovens accountability system

(1999), who conceptualized disasters as 'crises socially constructed' characterized as (a) being non-routine and destabilizing, (b) causing uncertainty and disorder, and (c) posing at least the spectre of social–cultural collapse (1999; drawn from pp. 22–30).

For its part, the upside of a United Nations (UN) definition of anything is that it is globally recognized (globally accepted is another matter). The downside is that many UN definitions appear to result from committees, where many interests and perspectives manifest. In 2009, the UN offered that a disaster is 'a serious disruption of the functioning of a community or a society involving widespread human, material, economic or environmental losses and impacts, which exceeds the ability of the affected community or society to cope using its own resources' (UNISDR, 2009, p. 9).[4]

For the purposes of this paper, we are going to largely follow the initial 2009 UN definition of a disaster but add in a good portion of the Oliver-Smith and Hoffman (1999) conceptualization, which yields the following:

> A disaster is a social crisis characterized by a sense of great uncertainty, significant disorder, or potential collapse in a community caused by a serious disruption in its normal functioning and involving widespread human, material, economic, or environmental losses and impacts that exceed the community's ability to cope using its own resources.[5]

The point of all this is that disasters result from the interaction between a hazard event (e.g. earthquake, hurricane or toxic release) and vulnerabilities. That is, no hazard event is automatically a disaster; a hazard must have human and/or environmental vulnerabilities to act upon to qualify as a disaster. Indeed, one of the classic formulations of disaster research is hazard × vulnerability = risk (see Blaikie et al., 2003), which highlights that the probability of a hazard event in a remote and unpopulated area (for example, a major earthquake in many parts of the Aleutians) carries little to no risk per se because its occurrence would not interact with significant human and/or environmental vulnerabilities. In contrast, the probability of exactly the same hazard event in a highly urbanized area (for example, in Central America) would carry great risk, especially if the area were characterized by deficient construction and ill-advised land-use patterns – which brings the discussion here back to DRR.

The problem with 'DRR' as a concept is again not the lack of definition, but rather their number. The UNISDR (2009, p. 9) has defined DRR as:

> The concept and practice of reducing disaster risks through systematic efforts to analyse and manage the causal factors of disasters, including through reduced exposure to hazards, lessened vulnerability of people and property, wise management of land and the

environment, and improved preparedness for adverse events.

While perfectly understandable from an organizational mission perspective, this broad DRR definition does not provide the kind of focus we need to address the 'four core questions' of the adapted Bovens framework for effective accountability. Therefore, we wish to emphasize *vulnerability reduction* in this policy domain and define DRR as:

> a set of coherent and systematic actions designed to reduce the physical, economic, social and environmental vulnerabilities of a given community or component part of a community to its most relevant hazard or hazards. More specifically, for a given community or component part of a community, DRR constitutes actions designed to reduce the probability that losses from its most likely hazard events would exceed the resources required for effective local coping.

5. Analysing DRR-accountability: more easily said than done

With the concepts of accountability, disaster and DRR now at least manageably defined for our purposes here, we may start addressing, as a prelude to the IPW paradigm, those four core questions posed by Bovens. As we will see, the answers are highly problematic because:

1. There are no established forums or 'principals' for actors as 'agents' to regularly explain and justify their DRR actions, which is exacerbated by the fact that (a) the major DRR deficiencies in most communities result not from decisions and actions per se but rather from non-decisions,[6] inaction and/or merely symbolic actions, and (b) the media, often advanced as a possible default forum for DRR-accountability, is highly flawed for that purpose because of its short attention span and difficulty in capturing and explaining complicated issues with deeply rooted and often historical causes.
2. Multiple actors, including in the private sector, are required for any substantive DRR achievements, which raises the 'who' or 'many hands' issue (see Thompson, 1980) for effective DRR-accountability. This tendency towards confusion on the 'who' or agent question for accountability is especially problematic in disaster contexts when a traumatized community is paying at least temporary attention to causal questions about risk, but has little or no background information or documentation on pre-event responsibilities.
3. It is not immediately clear 'what' is to be accounted for in the short term when DRR deficiencies or failings in a community may not be glaringly revealed for 50–100 or more years, which circles back to the 'who' and 'many hands' issues because the 'what' has to be associated with someone or some organization as agent for a DRR-accountability system to function.
4. There are remarkably few political and economic leaders who feel any 'sense of obligation' to explain themselves on DRR issues except in the aftermath of a disaster, at which time rather sterile and self-serving blame management strategies ('spin') seem to take precedence over any forthright attempts at identifying disaster loss causes and clarifying DRR-accountabilities (usually plural in fact).

These interconnected problems of forums for principals and agents, their identity, the substance of accountability and sense of obligation are well illustrated, both negatively and positively, with both the Haiti and Chile earthquake events of 2010.

6. The 2010 Haiti and Chile events: preliminary accountability lessons

The total death toll from the 12 January 2010 magnitude 7.3 earthquake off the coast of Haiti near the capital of Port-au-Prince will never be known with any exactitude, and the more official estimates of 230,000 killed may turn out to be low. From an accountability perspective, however, exact numbers do not matter. What matters is how an urban environment fuelled by an export-oriented industrial and agricultural model would evolve in such a way (Port-au-Prince had a population of less than a million as late as the mid-1980s) that extreme vulnerabilities were created – despite the fact that Haiti had historic earthquakes and well-documented hazard maps from both the Caribbean Disaster Emergency Response Agency and its own Faculté des Sciences of the Université D'Haiti.

In fact, highlighting the vulnerability creation of the last several decades in and around Port-au-Prince was a small number of well-engineered and well-constructed buildings that came through the earthquake essentially unscathed. So, what preliminary accountability lessons may be drawn from the Haiti catastrophe? First, a relatively small number of building owners and their design professionals saw their structures in long-term investment perspective, were aware of the area's earthquake risk and used appropriate anti-seismic provisions. Their design and construction decisions were based, however, on rational calculations and voluntary adherence to professional norms – not on adherence to any code or regulation, which leads to the second point.

Lacking in virtually all aspects of effective governance, Haiti, even the capital, had no record of land-use planning and, crucially for the 12 January earthquake, no record of an enforced building code. Buildings just 'went up' in the words of one observer, because there were no established location and construction permit and supervision processes. That is, from an accountability perspective, no development 'actors' in Haiti felt any obligation to pass any of their siting or building plans by any 'forum' (which in effect did not exist anyway), nor were there any negative formal or informal consequences for their not doing so. In effect, this was a precise multi-decade non-decision making and inaction recipe for creating the massive urban human vulnerability that only required a major (not even a great) earthquake to become a national catastrophe.

The 27 February 2010 magnitude 8.8 Chilean event (a great earthquake, releasing more than 170 times the energy of the Haiti event) and resulting tsunami present an interesting comparison with Haiti, but really more of a contrast. The earthquake and tsunami occurred in the final days of the outgoing left-centre Bachelet Administration and the incoming centre-right Piñera Administration.[7] The official death toll has been set at 521, with all the victims identified by name, but the Government of Chile has been careful not to separate those killed by the tsunami from those killed in the earthquake, perhaps because the tsunami deaths could easily be attributed to the highly public intra-governmental warning-alert system failure discussed below.

Chile has established sets of land-use regulations, but they are dispersed among different levels of government, as is their enforcement. The country has also adopted a series of building codes starting in 1972, including a code for seismic resistance in 2003. In a 2009 paper, Ernesto Cruz estimated that approximately 25 per cent of Chile's built environment, including infrastructure, housing and larger structures, was highly vulnerable. Cruz also estimated that only 2–3 per cent of the post-2003 built environment (mostly in the rural areas) did not conform to the new codes.

While hardly optimal, as evidenced by the death toll, at least 15 major building collapses and serious structural damage to infrastructure, the Chilean combination of land-use regulation and a series of progressively more strict and enforced building codes made the 27 February 2010 event a disaster rather than a nation-changing catastrophe. That earthquake event was, after all and to repeat, magnitude 8.8, the fifth largest earthquake ever recorded, and would have likely been a catastrophe without the DRR countermeasures already in place in Chile.

More specifically tsunami losses and their causes, however, were, and will continue to be, debated in Chile for some time to come, because there was a clear and publicly admitted communications and decision-making breakdown within the Chilean government, a story that was followed closely by the media in the weeks following the disaster. The core issue was a communication gap between the Chilean Navy's Hydrographic and Oceanographic Service ('SHOA' in Spanish) and the Ministry of the Interior's Office of National Emergencies ('ONEMI'), which led to the cancellation of a tsunami alert at 4:56 in the morning (the earthquake occurred at 3:34 in the morning). A series of tsunami waves, however, continued to hit the Chilean south-central coast and offshore islands, including the two largest – after the alert's cancellation.

From an accountability perspective, in the pre-2010 disaster period in Chile, development 'actors' faced both land-use regulations and building codes, and clearly took them into account when siting, designing and constructing buildings of all types. While further detailed research is needed on the exact nature and functioning of the 'forums' in Chile associated with these review capacities and how formal/informal consequences were factored in, the important fact is that such forums *existed* and influenced development and construction decision making. While sterner tests likely await when a large seismic event occurs closer to Santiago, the capital of 6 million people, the result of a reasonably successful accountability system prior to the 2010 event in Chile was modest life loss, in stark contrast to Haiti six weeks earlier.

With the Haiti and Chile case lessons in mind, and given the very real conceptual challenges to establishing effective DRR-accountability systems discussed previously, we may now turn towards the future. The question becomes, 'For DRR-accountability, what would the new Inducing Political Will paradigm entail in most countries, especially in the context of the four core questions drawn above from Bovens?' The answer is an accountability strategy that would empower DRR advocates to 'speak truth to power' – and to do so safely and effectively.

7. The IPW paradigm: a five-step strategy

7.1. Step 1

Conduct pristine investigations. One of the primordial problems in most hazard-prone communities is the existence, and then the enforcement, of a requirement that any development plan, land-use change or construction/reconstruction project be evaluated on risk criteria (natural, technological or in combination). The result in most countries is a hodgepodge of uncoordinated development or redevelopment projects where precious little is known (or at least documented) about their hazard vulnerabilities and therefore their risks. When a hazard event occurs and those latent risks become manifest, turning areas into major loss sites, accountability is thus rendered difficult by the lack of pre-event documentation and the 'many hands' problem.

The solution is to set up 'DRR councils' as legally and functionally autonomous bodies from the local to the national levels free from conflicts of interest and with independent financing and adequate review capabilities to evaluate the risk components of all construction, development and redevelopment projects.[8] These reviews must be of unassailable ('pristine') quality so as to render them invulnerable to effective technical challenges in the courts and/or the media. These reviews would then be made publicly, permanently and freely available so as to establish a baseline risk assessment 'record' of all projects. That is, in terms of the original Bovens questions, publicly establishing 'DRR councils' would resolve much of the 'forum' problem for effective DRR-accountability systems.

7.2. Step 2

Speak truth to power. It is not enough to establish a publicly available risk assessment baseline for all construction, development and redevelopment projects. Equally important is to establish a public record of having informed the appropriate decision makers (public and/or private) of those findings from the local to the national level. This record would have to include the responses, or in some cases the probable non-responses, of those decision makers, which establishes accountability not only for subsequent *actions* but also, and most importantly, for subsequent *inactions*. This step would cut straight to the heart of the very common phenomenon of risk acceptance non-decision making, where explicit calculations and discussion of risk do not even achieve place on the agenda of considerations. That is, publicly establishing 'having spoken truth to power' would resolve much of the 'obligation to explain and justify' problem for effective DRR-accountability systems, especially after the creation of DRR councils as outlined above in Step 1.

7.3. Step 3

'Lay it all out' – publicly. It is quite normal, if unfortunate, that the people who live and/or work in various developments and structures are not aware of their risks until a hazard event occurs and they or their families, friends or co-workers become victims. It is equally common that the elites who make the decisions to build in certain places and use certain design and construction practices do not live and/or work in those places; indeed, it is much more frequent that they live and work in markedly safer places. That is, the political and economic elites who are 'accepting' risk often do so for others and for future generations (they are in effect *projecting* risk onto others), not for themselves or their progeny. Moreover, those 'others and future generations' are almost always unaware of the risks that have been accepted on their behalf.

The solution is a corollary to both Step 1 and Step 2: The independent body of Step 1 assures that the people actually living and/or working in a particular place or structure are fully aware of the Step 2 risk assessment baseline record of that specific place and the subsequent political and economic elite non-decision making and actions/inactions about its risks.

7.4. Step 4

Create guilty knowledge and authority accountability. One of the most common blame management strategies of political authorities after a major disaster is 'we didn't know,' which is a combination of 'ignorance defence' and 'accountability shift' to a prior administration, government or even regime. While this is not unreasonable in some cases given long recurrence intervals for many major hazards, very few major loss sites are truly unknowable. The scientific and engineering knowledge about certain locales and certain types of structures is simply too deep in the 21st century for this type of strategy to be credible under any reasonable scrutiny.

The solution is to provide publicly, and in some ways theatrically (a press conference and web postings for example), each new administration or government with the catalogue of hazards, vulnerabilities and consequent risks in every part of every town, city and region of the country. Properly recorded, in one fell swoop this eliminates the ignorance defence while also tying the incoming administration or government to the non-decision making and actions/inactions of their predecessors, also reducing the effectiveness of the accountability shift ploy. While not perfect because it is difficult to hold current officeholders responsible for non-decisions and actions/inactions decades earlier, the fact is that most political elites have party identification and multi-generational 'lineages' to which they often lay claim – until a disaster occurs and they have to try to distance themselves from, or even disavow, those lineages. That is, publicly establishing 'guilty knowledge' would resolve much of the 'who' problem for effective DRR-accountability systems.

This hazard–vulnerability–risk assessment catalogue would then be matched, again publicly, with a set of programmes and projects to rectify the identified risks. The overall DRR result would be to publicly create guilty knowledge among the political and economic elites about the risks in their societies and what is or had been technically possible to have reduced them. That way, when a disaster strikes they would have, metaphorically, no place to hide politically and perhaps even legally. That is, publicly establishing a 'rectification plan' would resolve much of the 'what' problem for effective DRR-accountability systems.

7.5. Step 5

Prepare for the consequences – and a solution. Speaking truth to power with a broadly accessible public record and establishing DRR-accountability for elite non-decision making and actions/inactions will not be popular with the relevant elites. Professionals with in-country positions, companies or careers should expect a negative, if not a hostile, reaction from authorities and other elites, and so they must be prepared for pressures to withdraw or desist. A solution, however, would be for them to become part of a transnational alliance of 8–10 DRR experts who would be on call to come to their public defence, if necessary in person. Such a group could not only validate the more technical aspects of the analyses, but also attract additional media attention to the DRR issues and thus reinforce the shift to the IPW paradigm.

Looking ahead, and failing a shift to this new IPW paradigm that might force real attention to DRR in the 21st century, one cannot but be pessimistic about avoiding truly stunning losses that appear set to result from the massive vulnerability creation of the 20th century. That is, a continuation of the old MTWH paradigm would seem to guarantee that we will see not simply more disasters, but more outright nation-changing catastrophes.

Notes

1. Not generally well known, there is a substantial politics of disaster literature, and it emerged in essentially three generations or waves, the most recent major contribution being Boin et al. (2009). See also the work of Olson and Gawronski (2010), which catalogues the literature back to 1925.
2. This paper was being finalized when the BP-Gulf of Mexico oil well blowout and subsequent petroleum leakage occurred, which raised and continues to raise fundamental accountability questions within the US Government and between that government and the private sector. For reasons of time and space, however, that case will have to be explored in a separate paper.
3. These cases should provide rich comparisons or, more accurately, contrasts. The senior author here has made a preliminary effort with www.forbes.com/2010/04/05/earthquake-infrastructure-haiti-chile-opinions-contributors-richard-s-olson.html.
4. This definition has its own pedigree as a more flexible (i.e. less restrictive) version of a much earlier and more severe definition by disaster research pioneer Charles Fritz (1961, p. 655).
5. In some cases, for example in small countries such as Samoa, the affected 'community' may be the entire national society, so we are simplifying here by eliminating the 'society' term.
6. The concept 'non-decisions' has a specific meaning in political science, where it refers to issue suppression or the power to keep issues off public, institutional and decision agendas. That is, non-decision making is not in fact 'deciding not to decide', which leaves a documentable trace, but rather the power to keep a troubling issue from even achieving an agenda position where 'deciding not to decide' is an option.
7. Chilean seismicity is much greater than that of Haiti, and a Chilean aphorism holds '*cada presidente tiene su terremoto*' ('every president has his earthquake'). Coming during the interregnum between Bachelet and Piñera, the 27 February earthquake could be said to have caught two presidents.
8. One potential model would be the independent Electoral Commissions found in various Latin American countries in particular.

References

Ackerman, J. M., 2005. *Social Accountability in the Public Sector: A Conceptual Discussion*. Social Development Papers, no. 82, The World Bank, Washington, DC, March.

Behn, R. D., 2001. *Rethinking Democratic Accountability*. Brookings Institution Press, Washington, DC.

Blaikie, P., Cannon, T., Davis, I. and Wisner, B., 2003. *At Risk: Natural Hazards, People's Vulnerability and Disasters*. Routledge, New York, NY.

Boin, R. A., McConnell, A. and 't Hart, P., 2009. *Governing after Crisis: The Politics of Investigation, Accountability, and Learning*. Cambridge University Press, New York, NY.

Bovens, M., 2005. Public accountability. *The Oxford Handbook of Public Management*, E. Ferlie, L. E. Lynne Jr and C. Pollitt (eds). Oxford University Press, New York, NY.

Bovens, M., 2006. *Analysing and Assessing Public Accountability: A Conceptual Framework*. European Governance Papers C-0601, 16 January, www.connex-network.org/eurogov/.

Bovens, M., 2007. Analysing and assessing accountability: a conceptual framework. *European Law Journal*, 13(4). 447–468.

Cruz, E., 2009. Present status of earthquake preparedness activities in Chile. Paper presented at the *Workshop on Making People Part of the Solution*, Istanbul, Turkey, 5–7 July.

Dubnick, M. J., 2002. Seeking salvation for accountability. Paper presented at the 2002 *Annual Meetings of the American Political Science Association*.

Fritz, C., 1961. Disaster. *Contemporary Social Problems*, R. Merton and R. Nisbet (eds). Harcourt, New York, NY.

Koppell, J., 2005. Pathologies of accountability: ICANN and the challenge of multiple accountabilities disorder. *Public Administration Review*, 65(1). 94–108.

Manin, B., Przeworski, A. and Stokes, S., 1999. Introduction. *Democracy, Accountability, and Representation*, A. Przeworski, S. Stokes and B. Manin (eds). Cambridge University Press, New York, NY.

Moreno, E., Crisp, B. F. and Shugart, M. S., 2003. The accountability deficit in Latin America. *Democratic Accountability in Latin America*, S. Mainwaring and C. Welna (eds). Oxford University Press, New York, NY. 79–132.

Mulgan, R., 2000. 'Accountability': an ever expanding concept? *Public Administration*, 78(3). 555–573.

Mulgan, R., 2003. *Holding Power to Account: Accountability in Modern Democracies*. Palgrave, New York, NY.

O'Connell, L., 2005. Program accountability as an emergent property: the role of stakeholders in a program's field. *Public Administration Review* 65(1). 85–93.

Oliver-Smith, A. and Hoffman, S. M. (eds), 1999. *The Angry Earth: Disaster in Anthropological Perspective*. Routledge, New York, NY.

Olson, R. S. and Gawronski, V. T., 2010. From natural disaster to political crisis: a '5C + A' framework for analysis. *International Studies Perspectives*, 11(3). 205–221.

Paul, S., 1992. Accountability in public services: exit, voice, and control. *World Development*, 20–7. 1047–1060.

Perry, R. W. and Quarantelli, E. L. (eds), 2004. *What Is A Disaster? More Perspectives*. Xlibris, Philadelphia, PA.

Quarantelli, E. L. (ed.), 1998. *What Is A Disaster? Perspectives on the Question*. Routledge, New York, NY.

Strøm, K., 2000. Delegation and accountability in parliamentary democracies. *European Journal of Political Research*, 37(3). 261–289.

Strøm, K., 2003. Parliamentary democracy and delegation. *Delegation and Accountability in Parliamentary Democracies*, K. Strøm, W. C. Müller and T. Bergman (eds). Oxford University Press, New York, NY.

Thompson, D. F., 1980. Moral responsibility of public officials: the problem of many hands. *American Political Science Review*, 74(4). 905–916.

United Nations International Strategy for Disaster Reduction (UNISDR), 2009. *Terminology on Disaster Risk Reduction*. UNISDR, Geneva, Switzerland.

ENVIRONMENTAL HAZARDS
Human and Policy Dimensions

Progress in natural hazard risk reduction: What hath development wrought?

STEPHEN O. BENDER*

Organization of American States (retired), 4936 Fairbank Road, Tilghman, MD 21671, USA

The development–disaster risk management agenda has been shaped over the last 25 years by development policies and practices that have isolated lesser developed countries' development agenda from dealing with risk to natural hazards, by intentional actions to create a theory and practice of disaster risk management alongside other cross-cutting issues, by attempting to nurture emergency management in the context of disaster risk management and by fostering competition for resources. Sovereign states, multilateral development banks and the international development community should collaborate in shifting paradigms to: consider all development actions as initiatives to reduce risk; separate emergency management policy and practice from disaster risk management; fold disaster risk management and climate change adaptation into development planning and lending processes so as to address risk to natural hazards; promote hazard, vulnerability and risk information as a public good; and insist on accountability and responsibility to natural hazard risk all along the development continuum.

Keywords: climate change adaptation; development; disasters; emergency management; natural hazards; planning; risk

1. Introduction

By 1984 the theme of integrated regional development was established in theory and practice in the Americas. In great measure the international development community was preparing and shaping the economic and social sector destiny and the built environment of developing countries under the title of development with then some 20 years of experience since the beginning of the Alliance for Progress in the 1960s.

The theme of disaster mitigation (for the purposes of this paper referred to as disaster risk management, a term that includes emergency management) as part of integrated regional development was introduced about 26 years ago and discussed in the first Ocho Rios Conference (Bender, 1984).

Choosing now as then, the theme of development and disaster risk management provides an opportunity to look at disaster risk management in a much needed broader context than death, destruction and disruption. It also affords an opportunity to look at the shape, structure and function of development with an emphasis on risk to life, infrastructure and livelihoods.

The thinking in 1984 on development and disaster risk management is still relevant, but things happen, and happenings have altered the development agenda, notwithstanding that development by definition has been difficult to define. The following discussion is about what institutions need to do to carry out the development–disaster risk management agenda.

2. Six happenings shaping the development–disaster risk management continuum

At least six happenings (presented in roughly chronological order) have and continue to

■ *Email: baybender@verizon.net

doi:10.3763/ehaz.2011.0008 ISSN: 1747-7891 (print), 1878-0059 (online) www.earthscan.co.uk/journals/ehaz

The Emerging View

... during the remaining years of this century and beyond, development theory and practice will be shaped to an ever-increasing degree by the timely response to natural and human events, many of which will cause strife, conflict, and disasters. This is in contrast to previous shaping forces such as the availability of broad expanses of uninhabited or little developed forests, valleys, and plains; the creation of entire new cities built around the extraction or transformation of a single resource; and the single purpose extension of infrastructure networks to colonize remaining large tracts of land. Natural and human events in the future will dictate the location, beneficiaries, affected sectors, and providers of financial support, often without the availability of alternatives in the traditional context of development.

...... (We) must avoid a situation where disaster relief becomes synonymous with development assistance, and where post-disaster reconstruction depletes scarce resources otherwise destined for new investments. We must be vigilant over institutional divisions between disaster management and development assistance so as not to create a void in policies, programs and projects which address the utilization of resources in the face of ever-increasing demands and vulnerability, particularly of the poor (Bender, 1989).

dramatically shape the development and disaster risk management continuum.

1. *Since the 1980s a politically created and led disdain for public sector planning, and a following of private sector economic preferences and pressures*

In large part and in many places with political instigation, the public sector has laid aside development planning, particularly land-use planning, regulation and public infrastructure investment to guide growth and change, only to follow purely private economic initiatives.

2. *Since 1989 the post-Cold War a societal shift in tasks from primarily addressing wealth and income inequalities to addressing risk*

As Ulrich Beck noted in *Risk Society: Towards a New Modernity* (1992), following the Cold War the task of modern society has changed fundamentally from one characterized primarily by addressing inequalities of wealth and income (*although such inequalities remain*) to one of addressing risk from natural and man-made hazards *which cut across traditional inequalities and always across political, administrative, economic and social boundaries* (italics mine).

3. *Since the early 1990s the stove-piping of international development assistance*

Disaster risk management is an example of departmentalization or stove-piping of international development assistance, with the most recent example being climate change adaptation. Stove-piping of disaster risk management has been superseded by climate change adaptation programmes and was preceded by civil society/livelihood/governance programmes, which was preceded by gender programmes, which was preceded by environmental programmes, all of which now exist side by side with development programmes containing old line sectors such as agriculture, education, energy, health, housing, telecommunications, transportation, urban, and water and sanitation.

4. *In the Americas by the mid-1990s for developing countries, losses outstripping support*

The countries' cumulative declared economic losses since the 1960s due to natural disasters have grown beyond those same countries' cumulative receipt of non-reimbursable, non-military overseas development assistance. Economic and social infrastructure, destroyed or damaged by natural hazard events, was developed with still outstanding loans from multilateral development banks, bilateral development and non-governmental organization assistance programmes. Much of the remaining infrastructure is vulnerable.

5. *Since the early part of this century, the intentional transformation of disaster risk management into a sector operating among peers*

Already broadly addressed in the third point, the making of disaster risk management into a sector has been undertaken by competing professionals and institutional units and by political and overseas development agencies looking for new areas to support, all bolstered by a UN thematic decade. Disaster risk management is a key manifestation of sovereign states' and the international development community's presence that has come about since 1984, and it has not come about due to any one disaster event, technological breakthrough or development approach.

6. *Yearly global expenditures on arms and defence that exceed all UN specialized agency budgets and wish lists combined*
The dark side of the view 'Government (and overseas development assistance) by program: we are what we are willing to financially support' draws into focus the sums of resources that are not expenditures on development. Capital resources are not the issue as societies spend prodigious amounts on arms and wars that far exceed the combined yearly budgets and projected needs of specialized UN agencies that deal with all aspects of development including climate change adaptation and disaster risk management (including humanitarian assistance), and associated economic and social sectors.

3. The present state of things

1. *The relative decrease in deaths due to natural hazard events, but the ever-rising increase in affected population and economic losses, the decrease in access to natural resources and the ever-growing yet seldom met demand for humanitarian assistance*
There is evidence of progress on a portion of the mitigation agenda from 25 years ago. The progress that has been made is in emergency management – preparedness, monitoring, alert and response. In describing this progress many institutions and happenings are involved. The progress is, however, distinguished by its relative distancing of development and development's relationship from the issues of mitigation, prevention, risk reduction, rehabilitation, recovery and reconstruction.

2. *The emergence of repetitive loss, disaster-ridden sovereign states*
Most sovereign states highly exposed to natural hazards and impoverished sovereign states do not have and may not have for the foreseeable future the ability to effectively address the impact of natural hazard events without outside assistance. But the international humanitarian relief community and its partner and sister programmes dealing with recovery, reconstruction and most importantly development have now recognized that repetitive disasters and the underlying risk factors make continued effective emergency management questionable in light of donor fatigue and the anticipated natural hazard events (including those related to climate change, not to mention war, civil strife, forced migration, ethnic cleansing, terrorism and religious intolerance).[1] There is recognition that whether or not there has been an increase in the number and severity of events, there has been an increase in risk through development. Such declarations make note that vulnerability to natural hazard events has not been significantly reduced and that continuing, reiterated vulnerability issues remain from one hurricane season and one ENSO episode to the next.

3. *The competition between disaster risk management and climate change adaptation, and between these programmes and line sector, environmental and social programmes for 'development' resources*
Climate change is a unique set of atmospheric and hydrologic hazards with identified causal factors and is associated with sector-specific, geomorphologic-specific, population segment-specific impacts. It has consistently held the attention of the development community because of the clear linkage between human intervention and global warming, and by extension to climate change. On the other hand, the disaster risk management community chose to identify with 'disasters' and a self-proclaimed vicious cycle, and build on emergency management institutional structures. It often has failed to involve the owners and operators of vulnerable economic and social infrastructure and those populations related to the same. There is institutional competition between the disaster risk

management and climate change adaptation communities for funding support. This takes place even as governments seek to integrate the two through the identification of climate-related hazard risks, the design of specific risk reduction measures and improved and routinely used climate risk information with planners and other decision makers, and the use of disaster risk management as a climate change adaptation tool.

4. *The failure at a national level to identify specifically who and what is vulnerable to which natural hazard, why, what can be done about it, who pays and who benefits*

Sovereign states have been reluctant to report on implementation of the Hyogo Framework for Action in terms of its Expected Outcome. The international development community is reluctant to identify populations, and economic and social infrastructure vulnerable to natural hazards. While the analysis of a sovereign country's ability to financially recover from a disaster (financial risk, which is different from physical risk and economic risk) is at an incipient stage, typically little is known either by hazard or sector analysis why and how the vulnerability came about, what are the costs and benefits of lessening the vulnerability and most fundamentally who should pay and who would benefit from the risk reduction.

5. *The absence of risk management mechanisms such as land-use planning, zoning, codes, permits and insurance in development actions to define and meet acceptable levels of risk*

Land-use management including planning and zoning, and building codes and permits should draw attention to the challenges and opportunities for DRM, particularly of the development community. The manner in which societies occupy the landscape, bring about the built environment, and designate or otherwise allow for the use of space reflects without equal the core values of the society, the role of government in the name of its citizens and the rights of the individual. Environmental and natural resource management, insurance, finance and public investment, together with land-use management are the principal instruments of society in shaping risk levels and addressing vulnerability of the built environment. These instruments are sometimes politically contentious, particularly in a democracy, and they most quickly and definitively draw into play the broad array of development stakeholders from the public and private sectors.

6. *The increasing call for external incentives and subsidies to undertake risk reduction measures*

Even in the most capitalist, market-driven economies there is ample empirical evidence that the threat of loss is not sufficient to force action to reduce risk. The explanation for this reveals the following: (1) reducing possible losses to natural hazard events is relative to the short-term gains by shunning investment in risk reduction, (2) risk management decision making carries little or no accountability or responsibility, (3) for many individuals (particularly the poor) risk reduction is beyond their physical, political, technical and financial control, and (4) in all cases avoiding risk reduction constitutes a subsidy which the development community largely accepts with few or no plans for covering losses sector by sector.

4. The development–disaster risk reduction linkage: the way forward

1. *Development theory and practice must set aside completely the 'disaster cycle' paradigm, repair the damage from its legacy and embrace risk reduction through development for the poor, which is essential to meeting Millennium Development Goals*

There is still a degree of presence and influence of the 'disaster cycle' in disaster risk management. While less in vogue than at the outset of the International Decade for Natural Disaster Reduction, the disaster cycle represents the impact of natural phenomena in a closed system outside the purview of development processes with the implicit if not explicit entry point being the 'disaster' or natural hazard event, followed by a series of interventions leading back to preparations and monitoring/alert for the next 'disaster'. The discussion of climate change and development has made visible the role of development as both the present driver of constructed risk and the possible driver of risk reduction related to natural hazard events of all types in a way that disaster risk management never has been. If risk reduction does not become part and parcel of the development process, there appears to be little prospect for a country to diminish losses due to natural hazard events so as to reverse the growing impact on society.

Each proposed development action, sector by sector, must measure its contribution to the Millennium Development Goals as well as the Expected Outcome of the Hyogo Framework for Action and the UN Framework Convention on Climate Change and the UN Framework Convention on Control of Desertification.

2. *Emergency management policy and practice must be separated from disaster risk management, which must focus on risk reduction, yet emergency management must function as part of each economic and social sector, public or private*

There is a growing global need and an existing shortfall to provide humanitarian assistance in situations of all types, including natural disasters. Disaster risk management deals with exposure of people and things to hazards and ways to avoid catastrophic loss and lessen risk in the short as well as the long term; emergency management looks at the possible, probable and actual impact of hazard events on people and the means to alleviate their suffering and needs. The preferable option is to separate emergency management policy and practice from disaster risk management. While disasters and the declaration thereof denote dependency, as does underdevelopment, disaster risk management is not an extension of emergency management. Emergency management is not an initial or end action of risk reduction or development. But as the development agenda assumes responsibility for dealing with risk sector by sector, emergency management must function as part of each sector's approach to saving lives using its assets. The task is to clarify the role of emergency management as a professional discipline separate from risk management, and to identify the areas of coordination, collaboration and cooperation with disaster risk management in development.

3. *The question of disaster risk management or climate change adaptation as the priority is superseded by development encompassing both to achieve risk reduction to natural hazard events on a priority basis*

The issue is not whether climate change, climate variability and climate change adaptation are part of disaster risk management or whether disaster risk management is an instructive approach in dealing with climate events. The issue is that natural hazard events have posed, do pose and will continue to pose risks to people, economic and social development initiatives as well as existing infrastructure on which these depend. The current vulnerability to climate events is ever more apparent. Rather than call for the cessation of development budgets, climate change adaptation and disaster risk reduction must be seen as part and parcel of all development efforts, which in turn must be considered and evaluated for their climate change adaptation and disaster risk management content. The issue is not integration but recognition of risk management in development, which must focus more on the absolute conditions of vulnerability and less on the relative conditions of wealth. Natural hazard management cannot be separated from environmental management and sustainable development cannot take place in conditions of vulnerability. Economic development is not sustainable in conditions of vulnerability.

4. *The task is for society to charge the owners and operators of vulnerable social and economic infrastructure with accountability and responsibility for risk reduction*

All development stakeholders must consider disaster reduction and package it in achievable component parts whereby sovereign states actually become less likely to declare a disaster – that is needing outside assistance to control an internal emergency – for specific humanitarian needs (medical attention, search and rescue, shelter, food, medical supplies, water systems and sanitary sewers, etc.), for specific pieces of the social infrastructure (hospitals, schools, fire and safety, etc.) and for specific pieces of the economic infrastructure (bridges, airports, agricultural irrigation systems, IT, electrical energy generation and distribution, etc.).

Notwithstanding the calls by some for decentralization and increased planning for climate change adaptation and disaster risk management, national governments and international institutions are confronting issues of bureaucracy, lack of information technology, the need to focus on life safety and inefficiency. Public sector planning is out of fashion and lacking resources, experience and expertise. Other issues to be overcome include clarity of roles, overlapping assignments and responsibility, and lack of research, enabling legislation, risk communication, access to international technical assistance and funding support.

The consideration, much less the inclusion of risk reduction in development, is still in the early stages of recognition. Climate change adaptation and disaster risk management initiatives often exist in large measure isolated from one another, particularly at the national level. Those who identify themselves with one or other of these initiatives usually think of themselves as a community apart. Yet both groups have similar aims and share benefits from each other's actions. What is often lacking in the discussions of integrating climate change adaptation and disaster risk management into development is the presence of sectors in the structure and function of national development planning, and the interface of the countries with the larger international development assistance and lending community, which is dominated by sector approaches. Sovereign states and their counterparts are accustomed to negotiate, borrow/lend, disburse and evaluate through sector mechanisms. Defining and acting on sector vulnerability and inter-sector linkages is the current weakness but also the potential strength in reducing loss of infrastructure and protecting populations if each sector accommodates both emergency preparedness and risk reduction outlooks and abilities.

5. Sovereign states – the most influential factor in development and disaster risk reduction: the way forward

1. *Disaster risk management actions have to be created, implemented and evaluated in the context of national and local development planning and implementation*

There is convergence, duplication and divergence surrounding disaster risk management policies, programmes and projects undertaken by political and technical entities in the public and private sectors at the national and local levels. The result shapes part of the underlying risk factors referred to collectively as a culture of disaster by design, brought about principally through development actions. National priorities separate development goals from risk reduction, a declared responsibility of the state. International DRM assistance is often triggered by disaster declarations. And for the involved national and international agencies, disaster risk reduction efforts are often unsustainable. There is overall progress in disaster risk management which varies by sector, but for the most part in the national and local arenas there are no goals, measurable levels of achievement or coordination between sectors. For climate change adaptation, overall there is no momentum through development for disaster risk management (Bender, 2009).

2. *All development includes risks and subsidies, and the poor have a priority stake in publically supported risk reduction actions*

The outlays for reaching higher levels of resiliency of the public and private sectors alike to natural hazard events have been and continue to be presented as costs, investments, add-ons and generally something outside the expected cost to create the built environment, service or product. When outlays are not made to manage risk, risk itself becomes a subsidy with a potential hidden cost. The cost, sector by sector, is not only the loss to owners and operators of economic and social infrastructure if destroyed or damaged but also the cost to the poor when risk is left unabated and they are impacted through indirect and secondary costs. Sovereign states must manifestly become risk adverse on behalf of the poor. National commitments to United Nations conventions covering poverty reduction, risk reduction and environment, and regional and sub-regional acquired commitments must be used for risk reduction for the poor.

3. *Schools, health facilities and water systems must be the priority vulnerability reduction targets*
Prior to the International Decade for Natural Disaster Reduction and even now in some development sectors, focusing on specific hazards and specific types of buildings met with disdain. It is now clear that risk reduction does not come about unless and until specific hazard types and specific portions of the built environment, services or products are targeted for risk reduction. Focus should be on lessening the impact on schools, health-care facilities and places of public assembly as the priority.

4. *Natural hazard and public sector vulnerability and risk information must be presented as a public good on a no-cost access basis*
The public sector has not asked for, nor under present circumstances will it receive, sufficient natural hazard, vulnerability and risk information. Such information is now sometimes generated through sector development, social, environmental and disaster risk management project preparation. But this approach is not and will not be sufficient to systematically and comprehensively cover the existing risk management issues. Nor is there the prospect of funding the preparation of such information from national sector budgets. Those perhaps in most need of the information are those least likely to have the resources for its purchase. Such information must be defined as a public good and generated through programmes underwritten with internal and external resources sufficient to assure open and free access to be used by all in all development initiatives.

5. *All public sector risk management initiatives must concurrently address physical and economic as well as financial risk*
While there has been progress on dealing with financial risk, particularly of central governments in dealing with budget shortfalls for relief and recovery, the public sector has done little to analyse and prepare for dealing with the physical and economic risks associated with natural hazard events. Dealing with only financial exposure makes more visible the issue of who is at risk and who benefits. Each risk management action should initially consider the physical and economic as well as the financial risk components of the vulnerable economic and social infrastructure, population, service and product.

6. *National and sector development plans and actions must have clearly stated, measurable natural hazard risk reduction goals, objectives, means and beneficiaries*
Notwithstanding the clarity of individual approaches for addressing CCA and DRM in an integrated manner at national and local policy and planning levels, there are identified and notable gaps, obstacles and challenges to a seamless integration. Not surprisingly nor in any unexpected way, these limitations mirror those similar impediments of development policy, planning and implementation. To begin at the policy level, in most countries there is a lack of national consensus and a common, unified voice from national government on climate change adaptation and disaster risk management reflecting a lack of long-term strategy, legislation, leadership and effectiveness by named focal point institutions (Bender, 2009).

7. *Disaster reduction must first and foremost mean no dependency on external humanitarian assistance*
There is much to be said for sovereign states undertaking 'disaster risk reduction' in the truest sense that the specific needs generated by the impact of natural hazard events – medical attention, search and rescue, food and shelter, restoration of primary services, education, etc. – can be adequately handled by the government in concert with civil society organizations and the business community without depending on international support. Such national policy and action will help to clearly separate emergency management from disaster risk management to the benefit of both.

8. *Avoid individual and group actions, public or private, contributing to greater societal vulnerability*
In the much discussed subject of the tragedy of the commons, multiple individuals and groups of individuals in a society acting independently and solely and rationally consulting their own self-interest will ultimately destroy a shared limited resource even when it is clear that it is not in anyone's long-term interest for this to happen (paraphrased from Hardin, 1968). There is an analogous case of a natural hazard risk commons. Individuals, groups and government building component parts of economic and social infrastructure create risk as a subsidy in lieu of outlays to reach visible, acceptable levels of risk. Collectively they increase the overall risk of society to the point that the sovereign state will be unable to deal with the ensuing emergency when a natural hazard event strikes. This often prompts a disaster declaration and the use of development resources to attend to the ensuing losses and reconstruction.

9. *Appropriate land-use planning and environmental management must be at the core of natural hazard risk reduction at all geopolitical levels*
Messages and legacies from politicians that government should not interfere in private development have created a generation or more of development practitioners who expect government acquiescence to the private sector lead rather than private business compliance with government as a guiding force in development. Specific actions are needed to best use regulatory mechanisms that are and could be made available to deal with natural hazard risk – from land-use planning to built environment design and construction, building codes, inspections and occupancy requirements. Barely noted but of great importance in the Hyogo Framework for Action is the reference to the role land-use and environmental planning play in reducing risk, beginning at the municipal level, but also recognizing the role provincial/state/parish/canton-level government as well as the central government have to play. Such public sector initiatives, while counter to some political wishes, call into play consultation with civil society and other development participants to discuss and decide on risk creation and risk reduction.

6. Multilateral development banks: the way forward

1. *Multilateral development banks must recognize, accept and use all development programmes to manage natural hazard risk including climate change*
Moving beyond post-disaster events shaping the development agenda of borrowing countries, multilateral development banks should use development programmes to shape the financial, economic and physical risk reduction through development actions.

2. *Development lending and grants must not add incrementally to the vulnerability of society*
Development actions, particularly those actions supported by the multilateral development banks, must not create risk where none existed or add to unaddressed risk issues without such risk being identified, discussed and accepted by responsible parties.

3. *The current emphasis on financial risk transfer must be broadened to cover physical and economic risk, particularly of the poor and of public infrastructure on which not only the poor depend*
Multilateral development banks must become proactive players in risk reduction for and by their borrowing countries. They should complement the present emphasis on financial risk transfer with economic risk and physical risk reduction using all forms of development actions. They should move away from defining and benefiting from the risk commons (abhorring everybody's risk but protecting themselves from their specific risk). Such actions would also address cross-cutting issues such as governance, transparency, visibility, participation and environmental degradation.

4. *Promote a risk-averse position on the part of the borrower by insisting on risk assessments, and on insurance and other risk transfer mechanisms when financing economic and social infrastructure through multilateral and bilateral development assistance programmes*

In the area of risk transfer, multilateral development banks, bilateral agencies, governments and private risk management partners are experimenting with natural hazard event-driven (parametric) access to provide post-disaster capital as a complement to the traditional country access based on a disaster declaration. To date, neither approach has dealt in any depth with the issue of accompanying financial risk management with physical and economic risk mitigation as a prerequisite for qualifying for access. Multilateral development banks should promote if not demand borrowers to use private insurance market mechanisms to cover the risk of assets built with loans in lieu of the call for the multilateral development banks to accept technical responsibility for assets built with loan proceeds.

5. *Create, modify and otherwise adjust operation policy and procedures to fold disaster risk management, ex ante risk reduction and post-disaster recovery and reconstruction into the established development contexts*

Multilateral development banks have found a national demand for, political support for, and have answered a need for 'disaster' assistance to the extent that such assistance now constitutes a significant programme and portion of their lending activity while not supporting in a significant way risk reduction of 'development' initiatives. Disaster assistance is often prepared and implemented separate from development and separate from climate change adaptation, environmental and social programmes. Disaster risk management should be part of main line development processes.

7. International development assistance community: the way forward

1. *In policy and practice, fold all disaster risk management and climate change adaptation initiatives into development actions*

For the international development assistance community, disaster risk reduction investment and technical assistance (including response, recovery, reconstruction and mitigation) have become a separate but parallel track for lending, grants, technical assistance and institutional professional advancement.

But for the development community, the issue is not climate change adaptation, which is in essence managing the risk to atmospheric and hydrologic events. Rather, the issues are the state of emergency as the events take place on the one hand, and the vulnerability of the sectors and how climate change adaptation conditions national strategies to reduce risk on the other. Climate change adaptation approaches are a looking glass to options and opportunities that are beyond the present understanding, knowledge and experience of many of those who are experiencing the risk. Nor is the issue disaster risk management, which is managing risk to all types of natural hazards. Development initiatives have to make visible the specific levels of risk to natural hazards and accept or modify those levels in a responsible and accountable way. In particular, more work needs to be done on managing risk from climate extremes and discernable climate variability at the national and sub-national levels.

Stove-piped programmes that separate risk reduction from development should be dismantled, but emergency management should be defined and executed separate from disaster risk management in development. At hand is not mainstreaming risk management in development as if it were bringing something to development, but it is rather to identify and make visible the risk to natural hazards that is actually present in development actions throughout the sectors. With the direct involvement of multilateral development banks and sovereign states, the international development assistance community should redefine natural hazard risk reduction through development processes. Economic and social sector units in inter-governmental organizations and agencies must demand, support and participate in all facets of natural hazard analysis, vulnerability assessment and risk management. Regional and sub-regional inter-governmental organizations should support legislation, policy and operations reform as needed at the national and sub-national level so as to reduce vulnerability in accordance with the Expected Outcome of the Hyogo Framework for Action and the Priority Action 4 of reducing the underlying causes of risk (Bender, 2009).

2. *Recognize that risk reduction to natural hazard events is essential for development actions to meet the Millennium Development Goals*

The international development assistance community must define risk reduction objectives through development goals, including the Millennium Development Goals.

Economic and social sector units in these organizations and agencies should pursue the following functions together with national and local governments:

- Demand, support and participate in natural hazard phenomena research, data analysis and monitoring.
- Prepare vulnerability assessments of their capital stock and operations.
- Estimate cost–benefit ratios for leaving or reducing the level of risk.
- Prioritize risk reduction objectives by population group, sector, geopolitical unit, infrastructure component and hazard type.
- Bring the necessary human and financial resources to bear to achieve selected risk levels.

Regional and sub-regional inter-governmental organizations should support legislation, policy and operations reform as needed at the national and sub-national level to:

- assign responsibility and accountability, sector by sector, for the functions assigned above
- demand emergency preparedness and response plans be developed, embedded and sustained sector by sector
- define natural hazard risk reduction as a goal of development
- facilitate active participation of civil society in all its forms at all administrative levels
- measure sector by sector each proposed development action for its contribution to the Expected Outcome of the Hyogo Framework for Action as well as for its impacts on the Millennium Development Goals and
- provide a comprehensive monitoring and yearly reporting on implementation of the Hyogo Framework for Action.

3. *Support and contribute to making natural hazard information, including for climate change, a public good at no cost to be used in all development initiatives*

There is no reason to believe that development, social, environmental and disaster risk management projects will ever generate the natural hazard, vulnerability and risk information needed to systematically and comprehensively cover the existing risk management issues of public and private sector development. Such information, beginning with specific hazard type, must be seen as a public good and generated through programmes underwritten with resources sufficient to assure open and free access to be used by all in all development initiatives.

4. *While reducing the vulnerability of the few, risk reduction efforts must not add to the vulnerability of the many, thereby avoiding the tragedy of the commons*

While most mitigation is taking place at the settlement level, it is being done in the absence of a comprehensive sector or broad geopolitical scale. Risk reduction for one community may have an adverse impact on the risk of a neighbouring community if vulnerability and capacity assessments do not take into account natural and built environment systems and networks that link the communities. Effective natural hazard risk reduction will remain primarily local and community based, but hazard, vulnerability and risk analysis must be broader in scope to understand the origins of the risk, the role development plays and the options for its reduction.

5. *Recognize and collaborate in emergency management policy but practise it as a programme and discipline independent of disaster risk management*

As a major player in the provision of humanitarian assistance in natural disasters and in soliciting, allocating and distributing support in emergencies or immediate need, the international community should push forward an agenda to separate emergency management from DRM. But as the development agenda assumes responsibility for dealing with natural hazard risk, involved entities should encourage each sector to initiate its own integral emergency management functions as part of each sector's overall strategic plan to protect both the sector's users and its own assets and resources. There should be a clarification of the role of emergency management as a professional discipline separate from disaster risk reduction.

6. *Through risk reduction for the poor, speak justice in society to truth even as truth speaks to power*
Recent efforts of individuals, civil society organizations, foundations and the like reflect the use of their power of convocation and non-politically correct talk to make the case for a reduction of the vulnerability of the poor an issue of justice, even as it challenges the powers that be to reduce poverty.

8. Conclusion

It is no longer proper to consider 'disasters' (or emergencies) as a sector; it is now necessary to involve the owners and operators of vulnerable economic, social and natural resource infrastructure as well as the associated populations. The question of 'whose vulnerability', particularly when it involves the poor, must always be asked, and all development participants must be considered as protagonists in risk reduction.

Disaster risk management must become a demand-driven activity intrinsic to development whereby information, knowledge and expertise are sought and applied by populations to their economic and social well-being, and the protection of the natural environment and physical infrastructure on which they depend. Simultaneously, but distinctively, the same information, knowledge and expertise needs to be expanded and shared for emergency management (Bender, 2009).

Note

1. Complementary to emerging failed state indexes, there no doubt will be in the near future a Disaster Risk Reduction Failed State Index where a four-part evaluation based on empirical evidence will classify sovereign states as to their ability to deal with natural hazard exposure including climate change and its impacts: (1) UNISDR HFA-driven expected outcome reporting, (2) sovereign state natural hazard disaster declaration history, (3) international humanitarian assistance interventions (UN and regional, bilateral, CSO and NGO health, food and humanitarian assistance agencies NGOs), and (4) in place building codes, zoning and land-use restrictions directly related to specific natural hazards (location, severity and frequency).

References

Bender, S. O., 1984. Disaster mitigation as part of integrated regional development planning. Proceedings of the International Conference on Disaster Mitigation Program Implementation, Virginia Polytechnic Institute and State University, Ocho Rios, Jamaica, 12–16 November.

Bender, S. O., 1989. Disaster prevention and mitigation in Latin America and the Caribbean. *Colloquium on Disasters, Sustainability and Development, A Look to the 1990s*, A. Kreimer and M. Zador (eds), Environment Working Paper 23. World Bank, Washington, DC.

Bender, S., 2009. *Reflection and Analysis Surrounding the Commitments and Initiatives to Support the Implementation of the HFA from a Regional Perspective*. ISDR and OAS Regional Platform for Disaster Risk Reduction in the Americas, 10 April, 2009.

Hardin, G., 1968. The tragedy of the commons. *Science*, 162(3859). 1243–1248.

ENVIRONMENTAL HAZARDS
Human and Policy Dimensions

Reducing disaster risks 1980–2010: Some reflections and speculations

IAN DAVIS[1,2,*]

[1]Lund University Centre for Risk Analysis and Management (LUCRAM), Lund, Sweden
[2]Lund University, Box 117, St-221 00, Lund, Sweden

Through personal reflections on themes covering past experiences, the author considers selected experiences of disaster risk management (DRM) that span 30 years of DRM, and makes some speculations concerning the future, and ponders whether within the selected themes the future looks positive or negative. The observations begin with some memories of the Ocho Rios Conference: a multi disciplinary expedition to examine housing and hazards in Pakistan; issues concerning education and training; and finally the success of a massive reconstruction project of rural dwellings in Pakistan following the 2005 earthquake. Some key questions are considered: the value of community-based risk management; the reduction of short- or long-term risks; the protection of critical facilities; whose reality is being addressed? official neglect of education for DRM; the conditions needed to create highly effective safe reconstruction in rural Pakistan.

Keywords: chain of safety; cliff-top model; hazards; housing; lifelines; reconstruction; risk management; risk perception; training

Introduction

Reflection on the past:

> By three methods we may learn wisdom: First, by reflection, which is noblest; second, by imitation, which is easiest; and third by experience, which is the bitterest.

Speculation about the future:

> Study the past, if you would divine the future (Confucius, 551–479 BC).

In 2011, we have passed the halfway point in the Hyogo Framework for Action 2005–2015, and are already two-thirds of the way in the quest to achieve the millennium development goals (MDG) by 2015. So this theme issue provides a convenient staging-post to reflect on the past and speculate about the future. It is clearly impossible, in a brief paper, to review what has transpired in three decades of attempts to reduce risks, but it is possible to distil some personal reflections in an unashamedly subjective manner. Most, if not all, of the lessons I have noted are unoriginal and many are well known, but they are identified or repeated in this paper since they have all shaped my own understanding of disaster risk management (DRM).

The paper is based around four theme topics that bring together personal activities concerned with approaches to DRM. These reflections are structured around a pair of models: a chain of safety and a cliff-top model that pictures a certain past and an uncertain future.

Some years ago I developed the first of these models to assist with leading training courses on mitigation that likened DRM to a linked circular chain, each link representing a specific safety element. The model was adopted to emphasise that a chain is only as strong as its weakest link – to stress the importance of each safety measure, and that to reduce risks a broad and closely

*Email: ian.robert.davis@gmail.com

doi:10.3763/ehaz.2011.0009 ISSN: 1747-7891 (print), 1878-0059 (online) www.earthscan.co.uk/journals/ehaz

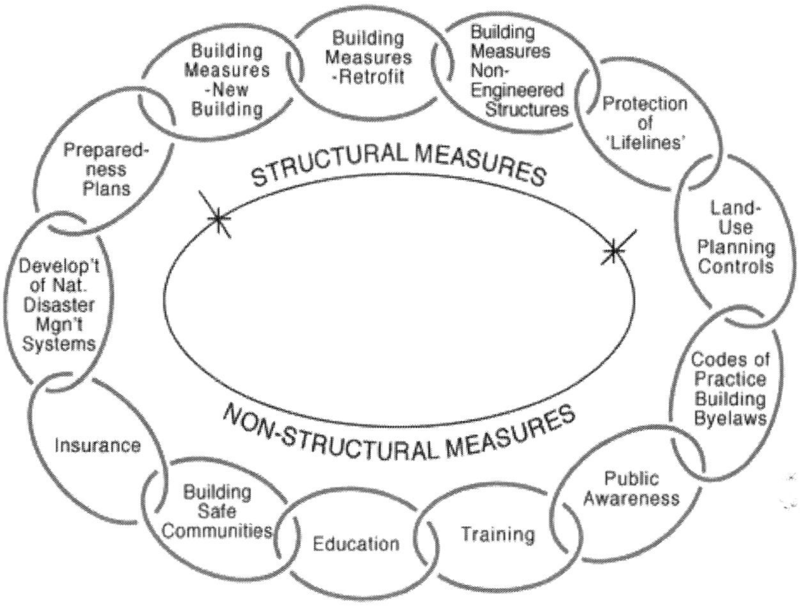

FIGURE 1 Chain model of integrated DRM (or safety from disaster)

integrated strategy is needed in each country, or local community, to bring together structural and non-structural approaches. Therefore, as a linking device for this paper I will use the chain model, and the various topics of this paper will be based on some of the individual links (Izadkhah and Davis, 2008).

Edward de Bono uses Figure 2 to visualize future forecasts, and I will adopt his model as my second visualization for this paper (De Bono, 1981). The present, as I write in November 2010, is the edge of the cliff face, from where we launch out into the unknown future. So the solid lines of certainty give way to dotted lines of uncertainty. One future scenario is in a positive upward direction, while the alternative falls away as a negative worst-case future. And horizontal dotted lines could also be drawn to represent anticipated continuity.

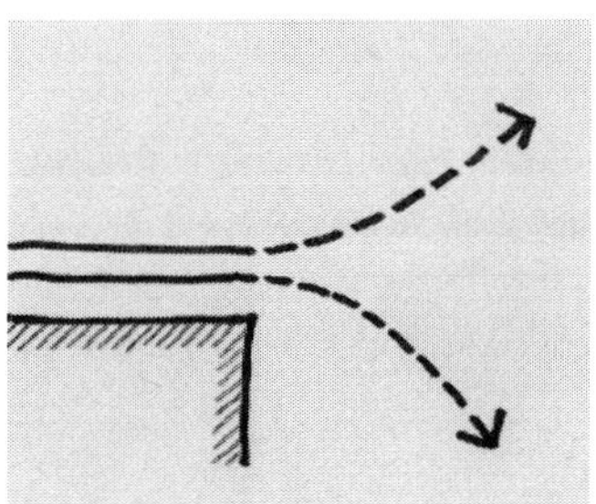

FIGURE 2 Future forecasts, the cliff-top model of future forecasts (a single positive and a negative option)
Source: De Bono, 1981.

In this paper I will look for some observed patterns of DRM drawn from my own history, whether upward or downward, that may inform the future management of disaster risks. The straight pair of lines that hover above the cliff in this model look back at the period 1980–2010 to a certain past (at least as far as memories serve), while the dotted lines look beyond my experiences to an uncertain future.

Theme 1

1.1. Cameos of risk management, observed during the 1984 Ocho Rios Conference

While many of the presentations in the 1984 Ocho Rios Conference are long forgotten, I retain vivid memories of three episodes about risk reduction that occurred during the conference. They concern risk reduction in relation to community action (where good progress has been made in subsequent years), conflicts in protecting against everyday vs longer-term risks (a persistent issue) and the protection of critical facilities (where progress has been depressingly slow).

1.2. Community-based risk reduction vs. official actions (See 'Building safe communities' chain link in Figure 1.)

My first recollection was of a public official's contempt of the value of local communities attempting to reduce disaster risks. I noted this conference exchange in the foreword to Andrew Maskrey's groundbreaking book of 1989 on a community-based approach to disaster mitigation, which was an expansion of his Ocho Rios Conference presentation. As Andrew gave his talk there was an angry reaction from my neighbour in the conference room, a Turkish government official, with responsibility for earthquake protection:

> ..... As we left the meeting he placed his hand on my shoulder and shook it vigorously-'What on earth' he asked, 'are these people, and that speaker's organisation doing? It is clearly the Peruvian Government's job to protect its citizens – not to leave it to the erratic performance of non-government agencies, who lack the necessary resources'. I responded with the rather unoriginal comment that governments in all countries are overstretched in terms of cash, manpower skills and expertise and should therefore only commit their resources to what communities are unable to do (Davis (foreword pp. v–vi) to Maskrey, 1989).

But now, 26 years later, it would be rare to find any government official, with the possible exception of those based in highly centralized countries, who would not recognize, or even extol, the importance of the community role in disaster protection (see my second theme for an elaboration on this topic as well as the paper by Maskrey in this volume, where the author considers this theme in detail) (Davis, 2004a, b, c).

1.3. Short- vs long-term protection

The second memory came during a bus ride to a conference field visit to Kingston, the capital of Jamaica. Keith Ford, our tour guide (contributing author in this volume; Maskrey, 1989), stopped the bus as we descended through a wooded steep ravine. He drew our attention to a concrete bridge as it crossed the ravine at right angles. The design of the concrete walls that served as bridge hand rails had resulted in a conflict on safety grounds between two Jamaican government agencies. Keith explained that the Disaster Protection Agency was concerned that the concrete walls that solidly enclosed the road would be likely to block the flow of the river and debris when water overtopped the road in flash flood conditions, thus posing a *long-term threat to safety*. Thus, the agency wanted to remove the side walls to prevent the bridge from becoming a dangerous natural dam. But the move to cut off the walls was strongly opposed by the road safety agency, who wanted to retain the concrete walls to provide safety to vehicles and pedestrians crossing the bridge as an *everyday threat to safety*.

In the years that followed, the issue of protecting against conflicting short-, frequent and long-term, infrequent risks has often surfaced, and I will return to the issue in the second theme of this paper.

1.4. Protection vs Neglect (See 'Protection of lifelines' chain link in Figure 1.)

The third recollection took place when our field visit reached Kingston. A city government official with responsibility for disaster preparedness gave us a presentation on urban protection. We were gathered in city offices with a fine view over the sea, with the airport in the far distance, joined to Kingston with an access road, set on a sand bar. Fred Krimgold asked the official how safe that road was from being inundated in the event of a hurricane storm surge. He expanded the question with the thought that there could be a situation where the airport could become inoperative following a disaster, just when it was most needed to bring in relief supplies, etc.

The official then responded with a confident assertion that there was no likelihood whatsoever of such a breach in the road in a future hurricane, since its level had been raised above anticipated storm surge levels. Well, just four years later, on 12 September 1988 Hurricane Gilbert struck Jamaica, causing widespread devastation, fortunately with minimal loss of lives, but including damage to the airport and a breach in the airport road, preventing the airport from operating until repairs could be made.

The issue was publicized widely as an alarm bell concerning the need for public officials throughout hazard-prone countries to provide enhanced protection for critical facilities from extreme hazard forces. But the cautionary lesson was not often heard or applied by decision makers in the years that followed. For example, the collapse of an access bridge to Port Island, sited just off the City of Kobe (Japan) in the 1995 earthquake, had devastating consequences. The main city hospital (another critical facility) is sited on the island, and before a temporary access bridge could be installed, casualties with critical injuries had to be transported to the hospital in small boats. But to compound the problem, the newly built hospital itself had suffered severe earthquake damage.

A decade later, the appalling destruction of 3,424 schools in the 2005 Pakistan earthquake, resulting in the death of 18,091 children (24 per cent of all earthquake deaths) and 853 teachers, was a further shocking example of the failure of the government to protect critical facilities or to control the corruption that almost certainly lay behind the destruction.

Five years later, in January 2010, further tragic evidence of 'lifeline failures' came in the Haiti earthquake, where 22 hospitals were seriously damaged and eight destroyed in the three most-affected regions. Then in Chile's 27 February 2010 earthquake, 79 out of 130 hospitals in the affected area were damaged, with a total loss of some 4,700 hospital beds. Many of these were near, and in some cases just metres away from, new hospitals that had been built to 'safe hospital' standards and suffered little or no damage (PAHO, 2010).

In Haiti there was also the loss of a whole range of vital governmental and UN facilities. These included buildings and their occupants with vital roles to play in the disaster aftermath: the UN Headquarters building, resulting in the death of the Head of Mission and key staff, and the destruction of the Presidential Palace, the Tax Office, the Ministry of Commerce and the Foreign Ministry. There were also massive losses to the education sector in Haiti, where an estimated 38,000 students and 1,300 school teachers died and more than 4,000 schools were damaged or destroyed (Bakody, 2010).

It is noteworthy that these examples of repeated failures in critical facilities span countries with extensive resources such as Japan, China and Chile, as well as very poor countries such as Pakistan and Haiti.

PROTECTION OF CRITICAL FACILITIES

First reflection:

Within the past 25 years there have been valuable campaigns by the International Strategy for Disaster Reduction (ISDR), the Pan-American Health Organization (PAHO) etc. to create and maintain safe schools and hospitals. While some success has been achieved, the protection of critical facilities remains an elusive quest to cover all vital resources: (airports, communication routes to critical facilities, key utilities and infrastructure, vital government offices, places of mass public assembly, cultural heritage etc.).

(The dotted line on Figure 2 presents a downward curve)

> *First speculation:*
>
> The failures of schools and vital government buildings in earthquakes in Pakistan, Sichuan-China and Haiti were more likely on account of corrupt building and an absence of adequate supervisory practice than to any lack of financial or technical resources to build safely. Therefore, the future protection of critical facilities will depend on the ability of countries to prioritize risk reduction efforts and tackle corruption, particularly in earthquake-prone countries (Green, 2005; Transparency International, 2008, 2009).
>
> *(The dotted line on Figure 2 presents a downward curve)*

Theme 2

2.1. The housing and hazards study in the Karakoram expedition to Pakistan, 1980

For my second theme I will go slightly further back from the 1984 Ocho Rios Conference to 1980, to note a further recollection that spans 30 years. In 1980, the UK-based Royal Geographical Society (RGS) celebrated its 150th anniversary by forming an international expedition to the mountainous Karakoram region of northern Pakistan. The expedition comprised a number of well-established scientific fields: geomorphology, geology, seismology, glaciology, etc. (Miller, 1982; 1984a, 1984b) (www.rgs.org/OurWork/Fieldwork+and+Expeditions/.../Pakistan/).

But one of the study themes of the expedition was certainly not established, and it is unlikely to have been studied in any interdisciplinary manner prior to this research work, and does not appear to have been systematically studied in a fully multi disciplinary approach during the intervening 30 years. The title of the area I had the privilege of leading was 'Housing and Hazards.' While housing in relation to natural hazards had been widely researched for at least half a century, and hazard research has had an even longer history, before our expedition it would appear that there had never been an attempt to bring together a multi disciplinary team, within a developing country, to examine the way in which communities, and their traditional dwellings and settlements, cope with or adapt to the threats posed by a variety of hazards: earthquakes, landslides, flooding, drought, etc. The research team we assembled comprised colleagues from Pakistan, Switzerland and the UK, and contained diverse academic disciplines: anthropology, archaeology, architecture, civil engineering, development studies, rural economics, human and physical geography, geology, physical planning and social studies.[1]

While the expedition glaciologists, seismologists, geomorphologists, etc., all using well tried and tested research methodologies moved into top gear from the day they arrived at the head of the Karakoram Highway, our team was trying to establish how to work together to answer a number of new research questions under the broad umbrella of 'Housing and Hazards'. Some of these topics have become common concerns 30 years later, but back in 1980 we found little to guide us in the available literature, or even from consultations with our eminent advisers. We wanted to know the most appropriate and reliable ways to determine the following:

- The seismic resistance of indigenous vernacular dwellings (Coburn, Spence et al., 1984).
- Any underlying causal factors that made these communities vulnerable or resilient (D'Sousa, 1984; Moughton, 1984).
- The perception of risk of local communities (Davis, 1984).
- Any practical actions they took to reduce risks or prepare for them in the design of their dwellings or in their social life (Hughes, 1984).
- The risks communities faced from natural hazards in relation to their everyday threats, such as health and livelihoods (D'Sousa 1984; Illi, 1984).
- In studying housing and hazard what contribution can one discipline bring to another?

What can an anthropologist bring to an engineer's grasp of the subject? Or an architect to a physical planner, etc. in researching these questions?

- And perhaps a pertinent question that was to emerge from our discussions with local residents, what did these local isolated, exceedingly poor communities, often living at a bare subsistence level, think of us and our concerns? What possible practical benefit could we bring to them? (For all the above references see Miller, 1984a, b.).

Gradually, we came to recognize the astonishing ability of the local families (perhaps better described as 'survival artists') to cope with the various hazards they faced (rock falls/erosion/floods/avalanches/earthquakes/droughts, etc.) that continually threatened not only their lives, but also their tenuous relationship with the landscape. However, 'cope' was the word we used to describe their toehold on the land. Only rarely did we see examples of families adapting their homes or living patterns to achieve better protection from natural hazards.

I vividly recall the interviews with heads of families within their cramped dwellings, and with a group of informative and thoroughly perplexed villagers who stood behind me while I was drawing their houses. The sketch was of marginal importance, but was deliberately drawn in a protracted manner to elicit valuable information concerning the way the occupants of the houses in my sketch perceived risks, rather differently from my expectations. Fortunately, many of the older residents spoke excellent English.

> When we questioned them why they paid so little regard to these risks they replied that in their eyes there were far greater worries than any of these hazards. In fact, one house owner shook his head in total mystification as to why we should come all the way from England to consider such a strange matter. Local priorities were more to do with their children's education (or lack of it), their health and the lack of medicine in the dispensary, the difficulty in selling their crops for a decent price, which is related to the problems of access to available markets given the very inaccessible terrain (Davis, in Miller, 1984a, b, p. 224).

Thus, the greatest personal value of the expedition came from what I was taught by these villagers, a key lesson to be of profound value for a full 30 years. While UN officials continually emphasize the need to mainstream DRM into every aspect of public life and grapple with ways to link DRR and the MDG, these villagers living in remote areas on a subsistence level of incomes were practising a holistic approach to risk assessment and management in relation to their traditions, culture and search for sustainable livelihoods.

The lessons from the Karakoram study were powerful and even more relevant today in 2010 than in 1980: vulnerability to short- or long-term natural hazards has always to be recognized within its natural context of the *everyday* livelihood, health and social concerns of affected communities. Just as within our affluent society, where long-term, logical action is often overwhelmed by short-term expediency, the economically poor but culturally rich families we met in the remote villages of northern Pakistan were preoccupied with immediate rather than more remote long-term threats. Expressed another way, DRM can never be isolated from the business of living; it has to be regarded as a key element in overall human development.

Our team arrived with lots of baggage, including some naive assumptions that we knew the risks families faced, and we also imagined that we knew the solutions to meet these needs. And reflecting with hindsight on the experience, it is clear that we were not prepared to revise our preoccupation with natural hazards in favour of considering their own risks, which they had clearly articulated. In the end, we gained greatly from our contact with these communities, but alas, to mix some metaphors, it was mainly 'one-way traffic' since our 'baggage' did not contain any of the assistance they urgently needed. When they asked us to provide tangible help to meet their needs or risks, we had nothing to offer.

Later, I concluded that undertaking research of this nature had implicit dangers of raising false expectations of forthcoming support with the families who were interviewed. And in the decades that followed, an answer was to emerge in the form of community-based disaster risk management (CBDRM). This was a diagnostic risk assessment set of tools that, at their best, combined community assessment leading to community actions to reduce the risks that they had helped identify (Davis and Murshed, 2006). (See 'Public awareness' chain link in Figure 1. However, the awareness in this case is from the public at-risk to decision makers.)

Following the 2004 South Asian Tsunami, a global evaluation of international assistance was undertaken. The evaluation team led by Claude de Ville de Goyet and Lezlie Moriniere commented on the need to let households assess their own needs following a disaster. Their Recommendation 6 is to 'Empower the affected individuals or families to assess and prioritise their own welfare needs by using cash subsidies whenever possible'. They elaborated with a radical proposal, that if accepted, would result in the sudden death of one of the much cherished disaster response 'industries' – that body of expatriate disaster assessors who have their travel bags packed by their front doors of their houses, confidently prepared to depart at short notice to assess any needs of any people, anywhere, and perhaps 'anyhow'...

> The need for thematic assessments would be considerably reduced if, when possible, the affected people were given the financial means to decide whether they want a better shelter, a boat, food or any other welfare item brought at high cost by expatriates. This approach would go a long way toward compliance with the Sphere principle of 'respecting the dignity of victims' in countries with active market economies... (de Ville de Goyet and Moriniere, 2006).

The decision as to *who* assesses needs, and *whose* needs are being met was the theme of Robert Chambers' influential 1997 study with the title *Whose reality counts, Putting the first last* (Chambers, 1997).

In the coming years, communities (or locally based groups) will increasingly have little alternative other than to assume direct responsibility for the assessment of the risks they face, as well as for the decisions concerning whether they must just live with the risks or take low-cost actions to protect themselves and their property. The reasons why this will happen are two-fold:

- First, in the poorest countries there will be severe resource limitations. As urbanization continues to put extreme pressure on already overcrowded cities, and as the population escalates in some of the most severely hazard-prone countries such as India, Bangladesh, Indonesia and Pakistan, it is unlikely that there will be sufficient local government capacity to allocate resources, or there will be sufficient professionally trained staff to provide technical support to a growing scope of vulnerability. Furthermore, the increasingly wide divide between the extremely poor and the expanding middle class and very rich will mean that engineers (and other professionals) will seek to engage with financially rewarding tasks. For example, there will never be significant money to be made from the design of risk reduction measures to 'non-engineered structures'. The failure of these buildings accounts for approximately 95 per cent of all fatalities caused by earthquake building losses.

  (See the 'Building measures: non-engineered structures' chain link in Figure 1.)
- Second, in many enlightened governments, there will be recognition that the *only* valid strategy to reduce risks will be via decentralization and devolution to local levels. This

change will probably give added impetus to CBDRM.

> WHOSE REALITY, WHOSE PERCEPTION OF RISKS?
>
> *Second reflection:*
>
> The dangers of paternalism. Whose needs and whose perceptions of risk are the basis of actions on the ground: the assessor's or the family 'at-risk'?
>
> *(The dotted line on Figure 2 presents a downward curve)*
>
> *Second speculation:*
>
> In countries with the highest poverty levels, local actions to assess, accept or reduce risks will become the norm, as local governments and professionals will often lack the inclination or capacity to deal with such escalating problems. While the omission of government support may stimulate local initiatives, it will deprive local communities of essential skills and resources and leave such groups unprotected.
>
> *(The dotted line on Figure 2 presents a downward curve)*

Theme 3

3.1. Training and education

(See 'Education' and 'training' links in Figure 1.)

My primary work over the past 38 years has been within an academic environment, undertaking research, leading undergraduate and master's courses, supervising PhD studies and leading training courses in DRM. The locations have been mainly based in British universities, but with teaching and research experience in Japan, Sweden, the USA, Canada, Mozambique, India and China.

A key observation concerns the cultural difficulties in teaching disaster mitigation to officials coming from an emergency management background. Participants coming from police, military or any of the emergency services tend to be well trained to follow a top-down 'command and control' working method. Thus, when the subject of risk reduction was discussed in management training courses they fully understood the concept of emergency preparedness, but when it came to longer-term developmental approaches to risk reduction they had a lot of re-learning to undertake. To 'get up to speed' they needed to be exposed to such issues as participatory decision making, capacity building, institution building, resolving ethical dilemmas, etc.

In the late 1980s a number of UN agencies, with a DRM role under the leadership of UNDP and the UN Department of Humanitarian Affairs, decided that there was an urgent need to develop a comprehensive training programme, primarily for their own staff, but to be widely available to government counterparts, NGOs, etc. It was called the *Disaster Management Training Programme* (DMTP). Excellent training materials were developed by the University of Wisconsin and INTERWORKS under the direction of Don Schramm and Paul Thompson, and these are still widely used in training courses. Gradually a wide diversity of training guides were developed, including emergency management, logistics, mitigation, development, reconstruction, etc. The programme involved Training of Trainers courses, Trainer and Trainee guidelines and detailed field testing, which led to further refinement in the programme.

At last, it seemed that the UN was taking disaster management training seriously, with a major step forward having been made. But alas, for reasons unknown, probably due to personality rivalries, agency politics or funding, the programme was cancelled in the mid-1990s. In 2007 an attempt was made to replace it with a new DRR training/capacity building initiative called *Capacity for Disaster Reduction Initiative* (CADRI) as a programme of the United Nations Development Programme's Bureau for Crisis Prevention and Recovery (UNDP/BCPR), www.unisdr.org/cadri/who-we-are.html).

However, little seems to have come of CADRI in the years since it was initiated. It has a far more limited scope than DMTP, being confined to risk reduction and early recovery, and has yet to replace DMTP as a vital tool in DRM.

While hazard-protective measures, such as building flood embankments, relate to all the opportunities and limits of *current* technology, a well-trained or well-educated person will benefit from today's learning experience in the *future*, perhaps in 30–40 years' time when the practice of risk management will then be radically different from the present. Thus, education and training are a vital investment in the future, unlike virtually any other protective strategy.

As the demands of DRM expand and as countries like India, Brazil, Indonesia and China grow in wealth, it will follow that they will wish to develop their own academic/professional development programmes in DRM, and in devising training courses.

An urgent need is to devise an international network of training and education courses where educational standards are established and externally validated and where teaching materials are freely shared. The general apathy and indifference of official bodies towards education is indicated by the fact that at no time in the intervening 29 years, when I have been leading disaster management training or master's courses, has any UN or international agency responded positively to repeated requests to convene a meeting of educational practitioners to form cooperative protocols, discuss the curricula the system needs when recruiting new staff and consider educational and training attainment standards.

However, in reviewing this paper, Andrew Maskrey has queried whether there is a need for a new generation of training materials given the profusion of training programmes now available throughout many countries, particularly in India and Latin America. He noted that young professionals can now easily get hold of information through the Internet, and that this easy availability 'must be feeding into education and training in a more organic way than training courses full of emergency managers'.

OFFICIAL NEGLECT OF TRAINING AND EDUCATION

Third reflection:

Education and training are perhaps the most important elements in any DRM strategy, since they (more than any other risk reduction measure) are a present investment, with benefits to be collected in the future. But this is not recognized in many quarters; therefore the subject remains a low-priority concern within United Nations agencies.

(The dotted line on Figure 2 presents a downward curve)

Third speculation:

Future disaster risk management education and training, at all levels, will be undertaken nationally, or regionally and will become less dependent on industrialized countries. It is a logical, developmental process for training and higher education to proceed with good cooperation between the institutes running programmes. But alas, the opposing competitive spirit between organizations and their staff can kill such cooperation, and in addition the self-interest of academic institutions within industrialized countries results in recruitment drives to secure students from developing countries, even if this reduces the development of national capacity.

(The dotted line on Figure 2 presents an upward curve)

Theme 4

4.1. Epilogue: the user-build rural housing reconstruction programme in Pakistan 2010 (See 'Building measures: non-engineered structures' chain link in Figure 1.)

Thirty years after the Karakoram expedition that I described in Theme 2, I made my first return visit to Pakistan to travel to the Azad Kashmir region of the North-West Frontier Province. My visit was to examine rural housing reconstruction following the devastating 2005 earthquake when over 70,000 lives were lost in an area of about 30,000 km^2. UN HABITAT had invited me to see the reconstructed rural housing they had helped facilitate in the area close to the epicentre of the earthquake in the villages surrounding Muzaffarabad (Davis and Mumtaz, 2010).

Having visited, researched, written and lectured about housing reconstruction following disasters since 1972 was delighted to observe perhaps the most impressive housing reconstruction programmes. For reasons given below, I regard this as a great achievement, particularly for the families who have built their own homes, but also for all the assisting groups who have made this happen.

As I left Islamabad, I reflected on what had transpired in the 30 years since I arrived in the city in 1980 to lead part of the RGS expedition (see Theme 2 above).

Negative news from Pakistan:

- The threats to Pakistan come from three directions: internal political insecurity, external threats from neighbouring countries and an explosive population growth rate. In 1980 the population of Pakistan was only 84 million, but in 2010 it is estimated to be over 181 million (the sixth largest country in the world), and there is an ominous projection that it will reach 335 million by 2050. In fact, half of the global population growth in the next 35 years is taking place in just six countries, including Pakistan.[2]

  Undoubtedly the death toll from the 2005 earthquake of 73,000 reflects the high urban and rural population density. The dense population also relates to the massive flood disaster that is occurring while I write, in what was formerly called the North-West Frontier province. In August 2010 unprecedented flooding has been taking place throughout Pakistan with over 1,800 deaths and over 20 million people have been caught up in this disaster, with millions displaced. The UN Secretary General has described the disaster as being more serious in its impact than the 2004 Tsunami.
- In 1980, Pakistan was a politically stable country. But by 2010, it had become one of the least stable countries in Asia, and this is undoubtedly increasing the vulnerability of the population. The following email message was posted on August 8 in a Caribbean Network of Disaster Practitioners, by an author who will remain anonymous:

I don't normally make comment but the unfolding events in Pakistan have stirred my professional passion. This is not the first time this area of Pakistan has been flooded, and it has even been flooded this badly before. It is the first duty of government to protect its citizens and too often this is interpreted as buying the latest military toys and weapons. Yet according to the US Embassy in Pakistan in the last three years the US has provided $3 billion of security and civilian assistance to Pakistan. The same article listed $780 million in military purchases from the US alone in the same time period. Simple question, how much did Pakistan spend on flood prevention or mitigation in the same period?

It's Pakistan this week, and unfortunately they are not alone in this misdirected effort to protect their citizens. With climate change, technology dependency and a greater complexity in our infrastructures is this decade the one in which nations finally focus on addressing the built and natural hazards to their people?

Lord Julian Hunt, a leading UK spokesman on climate change, has written that monsoon flooding is set to become a perennial concern in Pakistan in the future, yet Pakistan has entirely failed to prepare for such disasters (Hunt, 2010).

According to Transparency International (the body that monitors governmental corruption), – 60–70 per cent of the £640 million spent by the Pakistan Government's Federal Flooding Commission since 1977 has been embezzled (Transparency International, 2008).

The result has been a lack of investment in dams and flood defences, as well as a failure to prevent the destruction of forest cover. While flood warning systems are now in place in China, India and Bangladesh, it is reported that 'few, if any' flood warnings were given before this 2010 disaster (*The Week*, 2010).

Positive news from Pakistan:

- The development of seismic safety measures for indigenous rural housing, a subject that had only reached a rudimentary stage of understanding in 1980 had become a 'mainstream concern' of the Government of Pakistan, the World Bank, Jha et al. (2010) and UN-HABITAT. By 2010 a total of 463,243 seismic resistant houses had been built in just three and half years within a developing country with severe resource limitations facing severe border terrorism threats.
- The development of major housing reconstruction programmes that adopted a 'user-build' approach that had been pioneered by Charles Abrams, John F. C. Turner and Fred Cuny in the 1960s and 1970s seemed to have come of age with the building of an astonishing total of 392,000 houses. The trend to adopt a 'user-build' approach to reconstruction is a most encouraging development, since right up the Gujarat disaster of 2001 it appeared that the trend was for governments and the international finance institutions such as the World Bank to put their faith and money into massive contractor-based reconstruction projects.
- It was most impressive to witness the enthusiastic support of the Pakistan Government for the owner-build approach, since traditionally governments have often preferred to support contractor-based approaches that are tidy to manage, easy to finance with simple accounting procedures and can reward private sector interests who may support Governments. But with such macro scale projects there is a far greater potential for corruption than in the micro scale of owner-driven approaches. And it is particularly impressive for the 'owner-driven', 'bottom-up' approach to user-driven housing to have been so enthusiastically adopted by a military government. In similar situations in other countries, such military governments have been prone to support strong centralized 'command and control' policies.

SOCIAL AND PHYSICAL DEVELOPMENT THROUGH RECONSTRUCTION

Fourth reflection:

Reconstruction provided the opportunity to introduce a developmental, 'user-build' approach to traditional rural housing that incorporated seismic safety measures, with a linked training course for builders and house occupants. The programme was sponsored by the government, the World Bank and UN_HABITAT, with the active involvement of local surviving communities.

(The dotted line on Figure 2 presents an upward curve)

Fourth speculation:

Disaster reconstruction will remain the main opportunity to be grasped by assisting groups 'build back better' and 'build back safer'.

(The dotted line on Figure 2 presents an upward curve)

5. Conclusions

The reflections in this paper concern the:

1. priority need to protect critical facilities *(a negative future)*;
2. dangers of paternalism when working with vulnerable communities *(a negative future)*;
3. official neglect of the critical importance of education as an investment in the future *(a negative future)*;
4. value of user-build safe reconstruction, as community development *(a positive future)*.

The speculations in this paper concern the:

1. impact of corruption in creating and maintaining disaster risks *(a negative future)*;
2. future for very poor hazard-prone countries, where there is likely to be a lack of government capacity to attend to DRR *(a negative future)*;
3. continuation of the trend in higher education and training with national and regional programmes, replacing the need for trainees or students to travel abroad to industrialized countries for their studies *(a positive future)*;
4. user-build reconstruction is a major positive development in promoting safer and better building *(a positive future)*.

Therefore, in returning to the diagram introduced in Figure 2, my rather simplistic speculations indicate a mixed future (indicated in Figure 3), with three positive trends, indicated by upward curves, and five downward trends, indicated by downward curves. There are many examples of significant progress in DRM since 1984, but they are being outstripped by the relentless expansion in vulnerability caused by global pressures of population growth, urbanization, institutionalized corruption, globalization and the marginalization of poor communities. Thus, the development of accountable and transparent governance is a matter of primary importance.

In 1984 our conference was preoccupied with the losses and devastation from the 1976 Guatemala earthquake when 23,000 died. As I write this paper, 26 years later, news continues to trickle in concerning the plight of Haiti, which suffered almost 10 times the Guatemala death toll in the January 2010 earthquake.

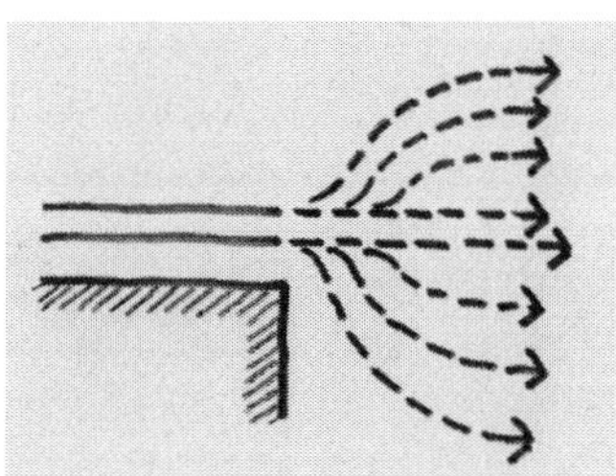

FIGURE 3 Future forecasts, the cliff-top model of future forecasts (multiple positive and negative options)
Source: De Bono, 1981.

Notes

1. The members of this group were: Ian Davis (Leader), Andrew Coburn, Israr-ud-Din; Richard Hughes, Dieter Illi, Cliff Moughton; David Nash, Frances D'Sousa and Robin Spence.
2. Half the world's population projected growth 2010–2050 is taking place in just six countries: Bangladesh, Indonesia, Nigeria, India, China and Pakistan (UN ECOSOC, 2009).

References

Bakody, J., 2010. *A Nationwide Call to Return to School Brings Hope to Children in Haiti*. UNICEF, 5 April, Haiti.

Chambers, R., 1997. *Whose Reality Counts, Putting the Last First*. ITDG Publishing, London.

Davis, I., 2004a. Position Paper: Social Vulnerability and Capacity Assessment (VCA). Prepared on behalf of the PROVENTION Consortium to Guide the Invited Seminar on the Development of Vulnerability and Capacity Assessment (VCA).

Davis, I., 2004b. *Community Based Disaster Risk Reduction: Critical Issues And Challenges Facing the International Community*. Report on Proceedings of the International Conference on Disaster Risk Reduction Kobe: The United Nations Centre for Regional Development (UNCRD).

Davis, I., 2004c. Progress in the analysis of social vulnerability and capacity. *Mapping Vulnerability: Disasters, Development and People*, G. Bankoff, G. Frerks and D. Hilhorst (eds). Earthscan, London. 128–144, Chapter 8.

Davis, I. and Murshed, Z., 2006. Critical Guidelines – Community Based Disaster Risk Management – (The Development of 'Process Indicators' to Measure the Effectiveness of Community Based Risk Reduction Measures). Asian Disaster Preparedness Centre (ADPC), Bangkok.

Davis, I. and Mumtaz, B., 2010. Report on the Rural Housing Reconstruction Programme (RHPP) Following the Pakistan Earthquake of 8 October 2005, UN-HABITAT, Islamabad.

De Bono, E., 1981. *Atlas of Management Thinking 'Future Forecasts'*. Penguin Books, Harmondsworth. 103–112.

De Ville de Goyet, C. and Moriniere, L., 2006. The role of needs assessment in the tsunami response; Let affected households assess their own needs. London: Tsunami Evaluation Coalition, ALNAP Secretariat, 64.

Green, P., 2005. Disasters by design. Corruption, construction and catastrophe. *British Journal of Criminology*, 45(4). 528–546.

Hunt, J., 2010. Deluges after the deluge. *The Guardian*, 24 August.

Izadkhah, Y. and Davis, I., 2008. Tsunami early warning system (EWS) and its integration within the chain of seismic safety. *Disaster Prevention and Management*, 17(2). 281–291.

Jha, A. K., Duyne, J. E., Phelps, P. M., Pittet, D. and Sena, S., 2010. *Safer Homes, Stronger Communities*. A Handbook for Reconstructing after Natural Disasters. Global Facility for Disaster Reduction and Recovery (GFDRR), The World Bank, Washington, DC.

Maskrey, A., 1989. *Disaster Mitigation, A Community Based Approach*. OXFAM, Oxford.

Miller, K., 1982. Living in a hazardous environment. *Continents in Collision*, The International Karakoram Project., George Philip, London. 113–126, Chapter 10.

Miller, K. (ed.), 1984a. The vulnerability and reduction of damage risk in small buildings subject to natural hazards. *The International Karakoram Project*, Vol. 1. Proceedings of the International Conference held at Quaid-i-Azam University, Islamabad, Pakistan, June 1980. Davis. 200–204, Chapter 27.

Miller, K. (ed.), 1984b. A critical review of the work method and findings of the Housing and Natural Hazards Group I. *The International Karakoram Project*, Vol. 2. Proceedings of the International Conference held at The Royal Geographical Society, London, September 1981. Davis. 200–204, Chapter 17; A. W. Coburn, R. E. Hughes, D. Illi, D. F. T. Nash and R. J. S. Spence. The construction and vulnerability to earthquakes of some building types in the northern areas of Pakistan, Chapter 18; R. E. Hughes. Yasin Valley: the analysis of geomorphology and building types, Chapter 19; F. D'Sousa. The socio-economic cost of planning for hazards: an analysis of Barkulti village, Yasin, Northern Pakistan, Chapter 20; C. Moughton. Barculti in the Yasin Valley: a study of traditional settlement form as a response to environmental hazard, Chapter 21; I. Davis. Analysis of recovery and reconstruction following the 1974 Patan earthquake, Chapter 22; D. Illi. The vulnerability to fire risk of some building types and traditional settlements with reference to the fire which destroyed Gayal-Kot in July 1981, Chapter 23.

Pan American Health Organization (PAHO), 2010. Press release *'Health Leaders Urge Action to Make Hospitals Safe in Disasters'*. PAHO, Washington, DC.

The Week, 2010. Report on Pakistan Flood Disaster. *The Week*, 21 August, 5.

Transparency International, 2008. *Preventing Corruption in Humanitarian Assistance*, D. Maxwell et al. Transparency International, Feinstein International Centre and Humanitarian Policy Group (HPG), Berlin.

Transparency International, 2009. *Global Corruption Barometer*. Transparency International, Berlin.

UN ECOSOC, 2009. Report by Population Division, UN Department of Economic and Social Affairs, NY.

ENVIRONMENTAL HAZARDS
Human and Policy Dimensions

Afterword

FRANKLIN J. MCDONALD[1,2,*]

[1]Disaster & Emergency Management Program, School of Administrative Studies, Room 204A, Atkinson Faculty of Liberal and Professional Studies, York University, 4700 Keele Street, Toronto, Ontario, Canada M3J 1P3
[2]Disaster Risk Reduction Centre, University of the West Indies, Mona, Jamaica

The previous papers provide stimulating and challenging material confirming that the fundamental technical tools and scientific concepts related to reducing the effects of hazard impacts on human society have been available to humanity for some time. However, the many allusions in this volume to the recent examples of significant losses (such as hurricane Mitch, the 2004 Indian Ocean tsunami, Hurricane Katrina and the 2010 Haiti earthquake) as well as the cases cited of regular less well reported, chronic smaller events also confirm that there are technical, social, regulatory, political, governance and accountability challenges still to be better understood and overcome. (This closing paper represents an active collaboration and synthesis of the views of the collaborating editors Bender, Davis, Krimgold and McDonald.)

Building the capacity to systematically assess and address hazard exposure, reduce risks and improve resilience, coping, adaptation and mitigation processes has been a recognized goal but one that has proven elusive to attain. Recent initiatives are referred to throughout the volume, and they include the UN system-supported International Decade for Natural Disaster Reduction (IDNDR), and the related International Strategy for Disaster Reduction (ISDR), and reflect an evolving conceptual framework related to an understanding that sustainable development (including the attainment of the millennium development goals – MDGs) will be impossible unless risk reduction and disaster mitigation are 'mainstreamed'.

The emerging recognition of climate-related risks as an additional factor adding significant complexity to the efforts at capacity and resilience building is noted in the earlier chapters. In addition, there is a propensity at many levels to create 'stovepipe'-like communities and constituencies in response to discrete development goals, multilateral agreements and sectoral or agency programming, which has the potential for impeding integrated holistic approaches to development challenges.

The preceding papers reflect the diversity of backgrounds and views of the contributing authors. Their broad underlying consensus on the need for more attention in the coming decades to be focused on the issues of political accountability, enlightened governance, genuine partnerships and combined local action is striking. Their views, particularly those of Olson et al., Collymore, Davis and Krimgold, on the cluster of issues related to political will, governance, weak accountability and the consequential erosion of public safety deserve careful consideration. Their insights and suggestions

■ ***E-mail: franklin.jmcd@gmail.com**

doi:10.3763/ehaz.2011.0010 © 2011 Earthscan ISSN: 1747-7891 (print), 1878-0059 (online) www.earthscan.co.uk/journals/ehaz

have the potential to influence strategies and follow up implementation activities in the context of the Hyogo Framework for Action (HFA) and similar initiatives at all levels.

Many of the chapters indicate or suggest that there is also a consensus, in hindsight, that the seminal Ocho Rios meeting could, or should, have more sharply and stridently called for stronger engagement of the social and behavioural sciences in disaster risk management issues in the context of development. While it is clear that the need for deeper engagement of the social sciences was recognized and advocated by the social scientists (and many others) present in Ocho Rios, one is left with the impression that there is wide agreement that the volume of that call could have been louder and bolder.

The contributions of Maskrey, Davis, Jones and Collymore suggest the need for cross-disciplinary actions and the deepening of genuine and functional partnerships to generate collaborative coping and adaptation action. Maskrey in particular has drawn attention to the positive benefits and need for local actions within a seamless framework of global, regional, national, subnational and local processes.

Several examples are cited of the positive experiences of combined actions linking global strategic aspirations such as the IDNDR, MDGs, the ISDR and the HFA to regional and national programming and linking them organically to innovative resource mobilization (Ford and Jones). However, the impression is conveyed that the examples of significant consistent success at the local (or sub-national) level such as Cuba and Bangladesh (from Maskrey) are the exceptions rather than the rule.

Bender and Krimgold raise provocative issues related to the current post Cold War development models and the need to take up again the reasons for some practitioners' concern that globally and locally vulnerability is increasing in absolute as well as relative terms. It is also felt that vulnerability is being 'built' systematically in many locations including megacities, rapidly urbanizing areas, coastal zones of low-lying states and small island developing states. It should be noted that, based on several recent events affecting developed states (such as Hurricane Katrina and the recent Italian Earthquake), this systemic growth of vulnerability is not confined to the developing world.

The 1984 articulation of the role of 'champions' as catalysts, robust advocates of long-term strategic action, conveyors and cross-fertilizers of good practice between isolated outliers of effort and as converters of risk reduction sceptics seems, from the perspective of this review, somewhat low key. This is somewhat surprising since the 1984 initiative was influenced by the late Fred Cuny, who was an active participant in the Ocho Rios pre-meeting processes, at the meeting and in its follow-up, as he was a recognized leader in international humanitarian crisis management. Perhaps there has been a cultural shift in the present era, when compared with the 1980s. In that earlier period, society created and enthusiastically followed its development heroes or champions (such as Ivan Illich, Paulo Freire or Fritz Schumacher), while in our current society we have become more suspicious, or even weary of charismatic leaders, and thus we turn away from them to place more trust in management rather than leadership, and relying on our frameworks and institutions.

However, the emerging consensus, appropriately reflected in the recent frameworks for risk reduction generated by the global initiatives such as the IDNDR, the ISDR/HFA and CDM, is that the overly techno-scientific approach alone is unlikely to lead to the resilience of economies, settlements and societies. Thus, given the increasing complexity of disaster risk management, coupled with the even faster expansion of vulnerability, the need for credible, passionate and articulate champions and leaders to give direction and inspiration is even more essential. But where are they? Where will such advocacy and leadership come from, given that it is highly unlikely that such leaders will emerge solely from the ranks of international bureaucrats or agency heads?

The 1984 Conference attracted stakeholders, experts, officials and the curious drawn from many backgrounds. It was, however, somewhat dominated by public officials, academics and those with a scientific or technical background. Their suggestion was that there is a need for a wider cast of players to address, strategize and implement development programmes and projects so as to fully embrace risk reduction (including natural disaster risks). The current volume restates emphatically the need for wider engagement and for deepening the process through the involvement of the private sector and for integration of effort at all levels. This should therefore be taken seriously.

1. Cross-cutting themes for future development

1.1. Role of the market and the private sector

During the past 25 years, development policy has been dominated by the Washington Consensus on neo-liberal economic thought. Pursuit of free-market economies has led to pervasive policies of privatization and deregulation. The benefits of the consequent globalization in terms of economic growth are well documented. These policies applied unwisely or indiscriminately may have had negative influences on some of the regulatory processes vital to public health and safety (e.g. building construction standards, zoning and land-use regulation). Some of these public safety processes have been weakened or stifled as a consequence of this era of 'free-market' development as there has been an increase (some would say an explosion) of vulnerable communities and run-away growth of disaster exposure, particularly in (but not confined to) developing countries. Regulation of physical development has too often been viewed as an unnecessary hindrance to rapid economic growth and job creation in disaster-prone areas.

Unfortunately, in the case of natural disasters, return periods may be so long that market values cannot and usually do not effectively reflect risk. Over the past decade a fascination with catastrophic risk insurance as a primary market response to natural disasters has also come into vogue at the international financial institutions. Various programmes of 'Cat Risk' have been implemented. However, they are poorly coupled to risk reduction and, in fact, have not proven to be an effective or efficient long-term means to effectively and comprehensively reduce disaster losses and the Cat Risk Tool should remain an area for operational and theoretical research.

The management and reduction of natural disaster risk require a renaissance of strategic, enlightened and balanced regulation in the public interest. Regulatory mechanisms must be developed that distribute the benefit of scientific understanding of hazards and vulnerability. At the same time, new innovative regulatory approaches must avoid exploitation by uncaring, corrupt or incompetent officials. The answer for disaster loss reduction is not deregulation but the evolution of innovative, transparent, just and competent regulation on behalf of a well-informed engaged civil society.

While the private sector has been viewed as the beneficiary of deregulation, it has also been the victim of growing disaster loss. In 1984, natural disaster mitigation was viewed as an almost exclusively public sector concern. Over the past 25 years, public/private partnerships have come to play an important role in disaster management. In free-market economies, the bulk of property and employment is in the private sector. Most decisions about investment and development are made in the private sector. The growing appreciation of common interests regarding disaster risk reduction by both the public and the private sectors offers hope that management of risk in the built environment may be possible.

The recent emergence and wide acceptance by key stakeholders – including financial institutions, insurers and multinationals of business continuity processes and risk management standards (led by the ISO and its members) – offer significant opportunities for the private sector to play a dynamic role in reducing risks over the next few decades. The potential benefits to the public sector and civil society groups of adaptation of the business continuity and continuity of operations practices already mainstreamed in many large corporation fall outside the scope of this volume but deserve to be given more attention as a trend to be encouraged and facilitated in the near future.

1.2. Role of education and capacity building

In 1984, post-graduate courses in disaster risk management were non-existent, and there were a mere handful of fledgling international training courses in disaster management. But now, almost 30 years later, we can see a global explosion of interest in the development of higher education in these fields with at least 12 courses being offered in UK universities and Andrew Maskrey has calculated that there are currently over 90 university courses on the subject in Latin America, and over 150 programmes in Canada and the USA.

Davis has suggested in his paper that it is possible to claim that education and training are the most important elements in any Disaster Risk Management (DRM) strategy, since they are a present-day investment in capacity building, with significant benefits to be collected in the future. But this reality remains a low-priority concern within United Nations agencies that have shown little or no interest in the DRM education sector as a focus for additional resources and to avoid wasteful duplication of effort. Bender and Krimgold insist that education and training in risk management must be part and parcel of ongoing, sector-specific professional preparation and practice.

1.3. The cultural/ethical/values dimension

One of the most obvious facets of moving forward relates to the challenging matter of integrating disaster risk reduction into development. This has been shaped over the last 25 years by changing development policies, priorities and practices that have increased vulnerability and that few would have anticipated nor called for in 1984. Some of these have, directly or indirectly, led to actions and results that have tended to isolate developing countries' development agenda from dealing with the risk from natural hazards, fostered intentional actions to create a theory and practice of disaster risk management alongside other cross-cutting development issues, attempted to nurture emergency (crisis) management in the context of disaster risk management, and fostered competition for resources.

The discussion, then, is not about integration of disaster risk reduction into development as an optional feature but rather that the current development agenda addresses in a wholly inadequate manner risk resulting from natural hazard exposure. In addition, other development issues such as adequate nutrition, education, health care, access to drinking water and sanitation, shelter and housing that most affect the world's impoverished are also not adequately addressed!

The variability, risk and uncertainties related to natural processes and hazard events (including climate variability and climate change) have been and will continue to be an integral part of development whether recognized and acted upon or not.

During the past quarter of a century, there has been a growing recognition that non-economic indicators of human development, quality of life, security and public safety (including the impact and risk of disasters) particularly focused on the poor and marginalized are an ethical as well as a technical, social, economic, environmental and political issue for many societies. Resolving issues such as safe built environments and other value or equity-linked development challenges – education, nutrition, health care, access to drinking water and sanitation, and affordable shelter – remain a problem in many jurisdictions in spite of the efforts linked to the MDGs.

In general, the solutions to the questions of how to address the ethical dimensions of risk and how to empower the marginalized and those victimized by unethical behaviour in order for them to become the primary actors in risk reduction remain elusive. The ethical issues – including moral hazard – related to natural hazard risk and disasters are fairly well known: settlements permitted or left to be constructed on totally unfit, dangerous sites in high-hazard areas; lack of enactment and enforcement of building codes and zoning regulations; corruption and cronyism in the award of contracts for buildings of public assembly and of high importance (schools, hospitals), theft and political manipulation of disaster assistance at all levels, etc. A host of countermeasures are routinely proposed to resolve these shortfalls: stiffer punishments for offenders, much more education for all stakeholders, public awareness and social responsibility being taught to children at all levels of schooling, programme and project monitoring and audits, policy and political reform, strengthening civil society institutions, transparency, etc. These have their place and value, and must be seen in a broader context of reform and modernizing of power relationships, promoting improved governance and improved public accountability with effective risk management as one of several outcomes.

Who should deploy risk management solutions on behalf of others; who will be empowered to seek their own risk reduction in the face of dominating opposition to change; and who should decide on acceptable levels of risk? In the end, with so many of the ethical failings being the products of poverty, the solutions are deeply embedded in the resolution of development challenge.

2. Critical issues

As a preamble, abundant facts and many descriptions, highlighted by the mass media, authorized by sovereign states and promoted by international entities through decades, declarations, conventions and goals, describe and decry the conditions contributing to the avoidable loss of human life, suffering and property losses due to natural hazard events. More detailed indexing and categorization, passing of resolutions and non-binding commitments have not by themselves brought about substantive reduction of disaster losses. Neither have they brought about substantive reductions of hunger, illiteracy, disease and human settlement deficiencies. Currently accepted approaches, including dealing with risk reduction in full or partial isolation from the development agenda, must change significantly if humanity is to attain sustainability targets and goals.

This theme issue has underlined four critical issues.

First, the significance and potential of international conferences, such as the 1984 Ocho Rios gathering in introducing young researcher/practitioners to their international colleagues, as well as enabling them to contribute their own insights into disaster risk management. An example of the process can be found in the attendance of young professionals Andrew Maskrey and Jeremy Collymore in 1984, who went on to accept global leadership roles in DRM, and have contributed papers to this issue.

Second, the elusive nature of progress within the dynamic environment of disaster risk reduction. Each paper reports on progress, but they also comment on the 'moving target' of complex, escalating problems that so often appear to outstrip available solutions. In 1984, there was extensive discussion concerning the appalling loss of life in the Guatemala earthquake of 1976 when 23,000 were killed. However, the backdrop to the writing of the theme issue has been the tragedy of the 2010 Haiti earthquake with some 220,000 deaths (almost 10 times the death toll of the Guatemala disaster). It should be noted that recent experiences following major recent catastrophic disaster impacts in the USA (Katrina), China, Chile, Italy, New Zealand and Pakistan (earthquakes) suggest that the escalation of losses (life and property) is not confined to developing countries.

Third, as disaster risks grow, enhanced resource flows follow. But with them come inevitable political constraints as highlighted in various papers. The critical challenge is to develop patterns of political accountability and transparency, particularly within fragile states and countries where effective governance is missing or limited, and to see greater exercise of pre-emptive anticipatory action to reduce human and economic losses. New conversations and dialogue mindful of balancing national sovereignty sensitivities with the brutal reality of future humanitarian crises and the lives at risk may be needed.

Fourth, a key focus of the 1984 conference was on how to make disaster mitigation 'stick' (or expressed in contemporary jargon – 'how to institutionalize DRR'). Despite the plethora of new agencies, global frameworks, academic programmes, etc. this challenge remains. It has in fact expanded as an urgent quest to link, or embed, risk management policies and practices into development (and the complementary chain of planning and regulatory processes); the management of everyday risks; and with programmes for poverty reduction. Just as the public health reformers of the 19th century sought, and largely succeeded – through political advocacy, public sensitization, awareness building and technical innovation – in integrating health concerns with building, sanitation, education, etc., today's 21st-century advocates of effective disaster risk management face similar, yet still unresolved, challenges.

We also, as a matter of priority, need to explore the appropriate role for the private sector and clarify the role of the market in supporting (economic) resilience building, and in promoting a culture of safety as a public good and investment in a safer future.

Several insights and suggestions emerge then from these chapters as to the way forward to a safer future. We need to shift our emphasis to address local, pragmatic but genuine partnerships; to continue to stimulate risk-related research and capacity building with deeper engagement of behavioural and social scientists. We need to target those responsible for development, not just disasters, and to seek to have ongoing strategic interventions in the political, investment planning and educational processes.